AF540996

INVENTIVE TECHNIQUES OF REFURBISHING DISCARDED TEXTILE MATERIAL

Compiled and Edited by
Dr. Rita Kant

Contributions of Research Scholars

Ms. Meeta Gawri
Ms. Sumita Sikka
Ms. Sugandha Saini
Ms. Harpreet Kaur
Ms. Preeti Alagh
Ms. Shweta Sharma

Gaurav Book Centre Pvt. Ltd.
Delhi

Publisher
Gaurav Book Centre Pvt. Ltd.
4832/24, Prahlad Lane, S-207
Ansari Road, Daryaganj, Delhi-110002
Phone: 011-43570976, 23278261
Email: gauravbookcentre@gmail.com

Published: 2015

ISBN: 978-93-83316-20-5

Laser Typesetting
JEE-VEE Graphics, Delhi

Price: 1795/-

Printed
Vikas Computers, Delhi

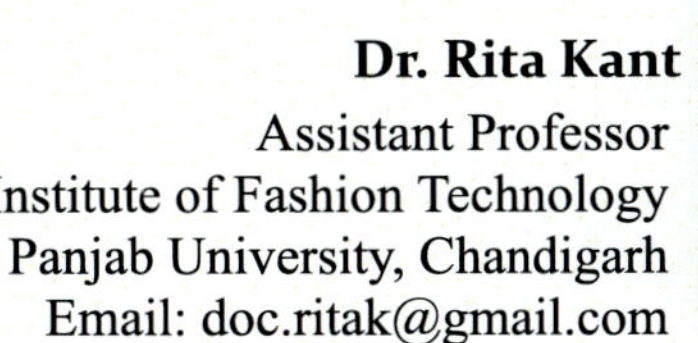

Dr. Rita Kant
Assistant Professor
Institute of Fashion Technology
Panjab University, Chandigarh
Email: doc.ritak@gmail.com

PREFACE

Textile industry is considered to be the most polluting and waste generating sectors. Over 80 billion garments are produced worldwide annually. According to the US Environmental Protection Agency, 13.1 million tonnes of textile are trashed every year. Fast changing lifestyle trends, premature fashion obsolescence and throwaway attitude of the modern consumer is resulting in large amounts of textile waste being dumped in landfills throughout the world.

Value can be added to a lot of this textile waste by converting it into useful products by simple techniques. It is with this intent that this work has been put together. Here is an effort to instill a sense of judicious and skillful use of resources by humanity for the sake of humanity; a humble message of "sustainable development".

This is an experiential project story narrated by research scholars; the works are ingenious and unique. All along, the objective is to invoke in the reader a skill to appreciate creativity; admire art works; value "Green Practices" in our daily lifestyles. The purpose is to motivate our readers to contribute their own; and while they indulge in this open ended art of self expression they get to practice values of diligence and consistency.

The products may not resemble commercial merchandise of an assembly line; but it is their flaws that make them artistic, ingenious, with a unique selling point (USP) of their own. Moreover this stress busting exercise of self discovery can become an income generating activity if one wishes it to be so.

In this time of Global Economic Prosperity resulting in wasteful consumerism, we wish our readers all the best in their endeavor to adopt "Green Practices".

Dr Rita Kant

Contents

Meta Gawri
Email: meetagawri@gmail.com
H.O.D- Fashion Marketing and Management
Assistant Professor, N.I.I.F.T., Mohali, Punjab

Department	Wastage causes
Sample Production	• Mistakes in design communication • Craftsmanship problem
Cutting Floor	• Wrong color or shade • Fabric faults • Less efficient marker
Sewing Floor	• Machinery problems • Faulty craftsmanship
Outsourcing	• Dyeing • Embroidery
Final Inspection	• Ironing problems • Measurement problems

Fig. 1 Departments of a garment industry.

RE-ENGINEERING WASTAGE FROM A CUTTING ROOM OF A GARMENT INDUSTRY

Recycling and reusing the valuable waste material can result in development of fantastic and usable products. Rather than putting these waste materials into the landfills, various innovative and creative ideas can be put together to make something new and useful.

The aim in the present work is to use the waste fabric and other left over trims to make useful products which can be sold with the garment as an accessory. Consequently the industry can earn profit and at the same time utilize resources to full capacity.

Various wastages occur in each department as garment industry is a labour oriented industry and human error is likely to occur at times.

Fig.2 Waste Bins with scraps of fabrics and trims.

I had this madness in me about "why to waste?" So I made an effort to empty out waste bins of a cutting department of a chosen garment unit. This waste comprised of scraps of fabric, pieces of laces, ribbons etc. I thought that each one of these could be engineered into acceptable and useful lifestyle product. I shared my project idea with a garment manufacturing unit. It was highly appreciated and executed under its "green practices". It was found to be easily worked upon and profitable.

After analysis of the waste material and much consumer deliberation based on the brand image the company wanted to promote , we decided that fabric jewellery to match the garment already being manufactured will have a USP and be a runway success achieving the goals of profit earning by the best utilization of discarded materials.

Fig. 3 Fabric left after cutting of garment parts

Fig. 4 Areas of fabric wasted after cutting

Delicate String Necklace

Dainty Headband

Flower Choker Necklace

Graceful Bow Clip

Product 1: Delicate String Necklace

Material Required:

1. Waste satin ribbon width ¾ inch
2. Pearl beads (approximately 50 pre-drilled pearls with a diameter of 6-millimeters)
3. A pair of 4 inch blade scissors
4. Polyster thread (a length of about 2 yards of size)
5. Cigarette lighter (smoking kills !! ☹)
6. Number 7 sewing needle
7. Glue gun or permanent fabric glue
8. A firm steel hook (20 mm size)

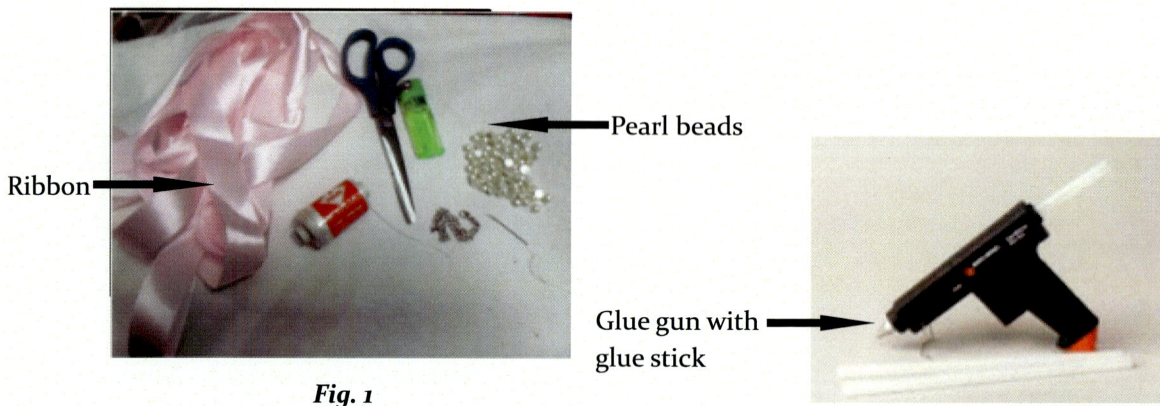

Fig. 1

Fig. 2

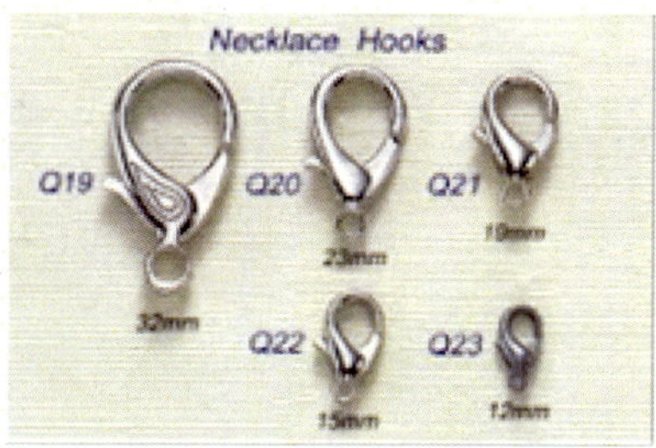

Fig. 3 Steel hooks

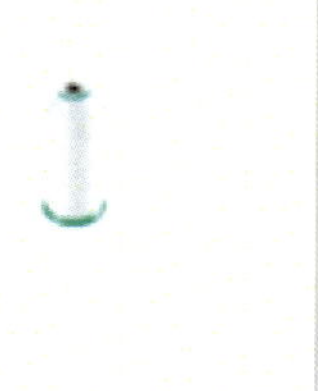

Fig. 4 Polyester thread

Fig. 5 Tweezer

Fig. 6 Cigarette Lighter

PROCESS

Step : 1

Fig. 1.1 Cut the ribbon into 1-2 inches in length.

Fig. 1.2 With a lighter burn the raw edges of the ribbon so that they fuse and do not allow the ribbon fabric to ravel.

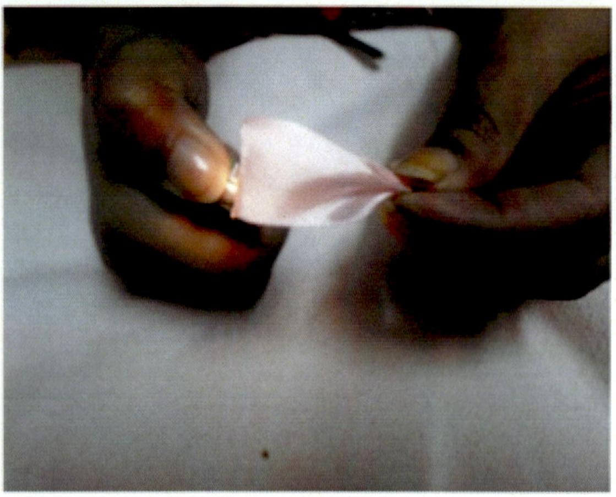

Fig. 1.3 Now hold the ribbon from one end and keep burning it till the edges are burnt slightly and the ribbon puffs up in a petal like shape.

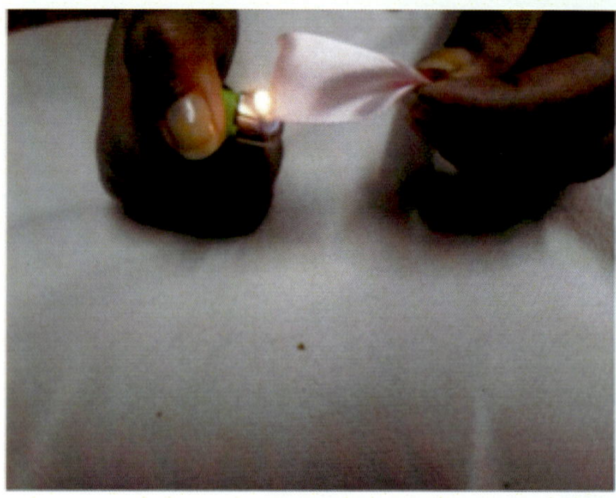

Fig. 1.4 Keep in mind not to over burn it, otherwise the ribbon may turn black.

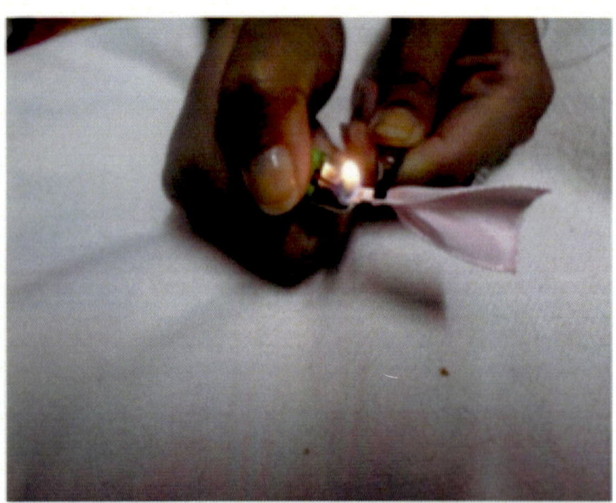

Fig. 1.5 Now burn the other end of the ribbon with a tweezer so as to fuse and fix it.

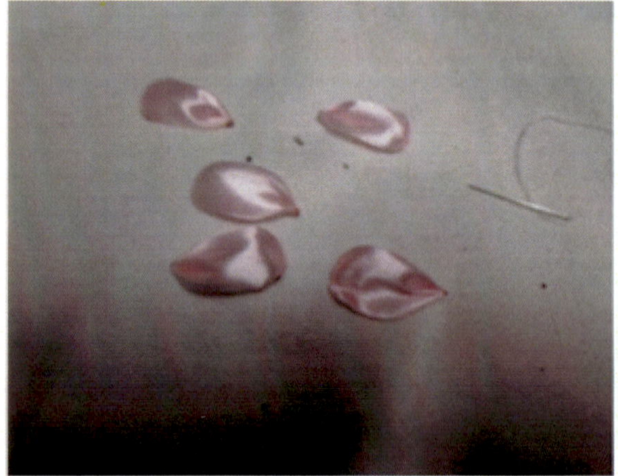

Fig. 1.6 Make as many petals as you want for your flower; here I have made five.

Step:2

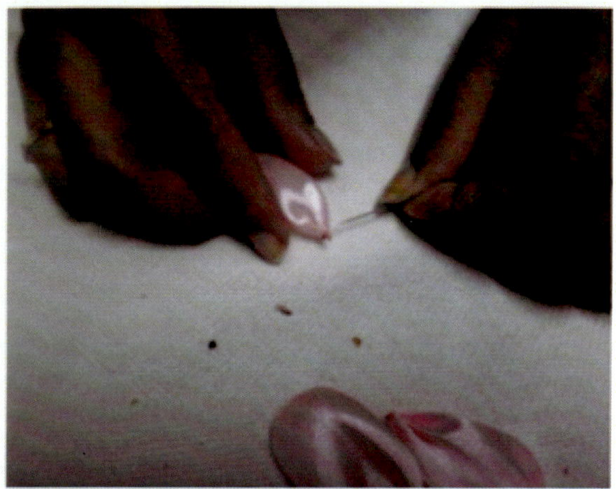

Fig. 2.1 Once you have managed to get a pretty petal shape brush out the extra burnt fibre using a tip of a needle.

Fig. 2.2 With the help of a needle and thread stitch and join the petals to make a flower shape.

Fig. 2.3 The flower will be made as shown above. You need to make three flowers for this project. You can use your innovation and can make variations in this flower by making it double or triple layered by the same technique.

step:3

Fig. 3.1 Take a nylon thread about 1 m long and fix a steel accessory hook at one meter end of it.

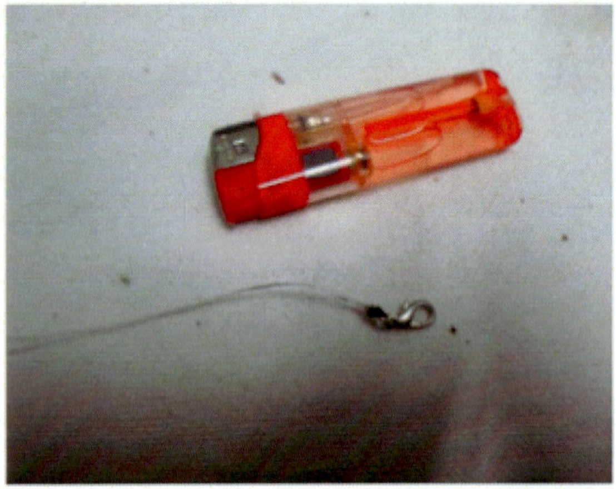

Fig. 3.2 You can attach the hook to the thread by the use of a lighter flame. This will fuse the ends of the thread by melting it in the heat.

Step:4

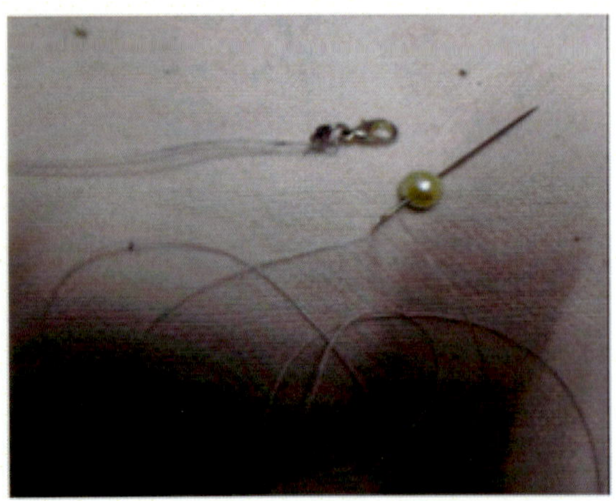

Fig. 4.1 Pass the other end of the thread through a needle and insert 25 pearl beads in this thread.

Fig. 4.2 After inserting 25 pearl beads insert one flower made in step 2.3.

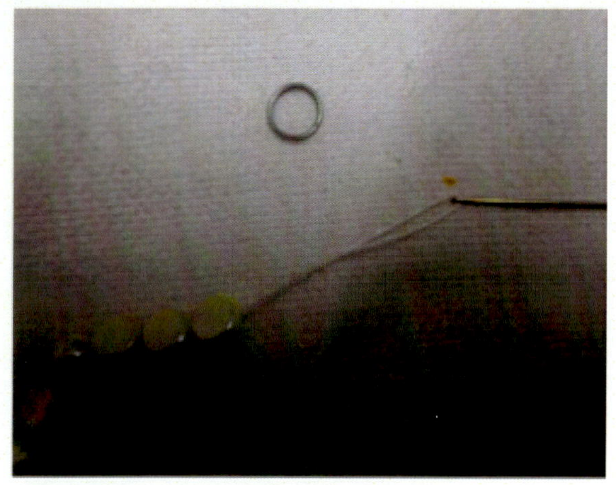

Fig. 4.3 As you reach the end of the thread fix the metal eye of the hook to it.

Fig. 4.4 Now your necklace is ready with the basic structure.

Step: 5

Fig. 5.1 With glue gun fix a pearl in the center of each flower.

Fig. 5.2 Follow this step carefully without spoiling the look of the flower.

Fig. 5.3 Complete all the flowers in this manner.

Fig. 5.4 Your Pearl Necklace is ready to be flaunted!!

Product 2: Dainty Headband

Material Required:

1. Waste satin ribbon in 3/4 inch width and in a color available
2. A pair of 4 inch blade scissors
3. Plastic headband with velvet finish
4. Cigarette lighter (smoking kills)
5. Glue gun or Feviquick
6. Tweezer
7. Crystal bead

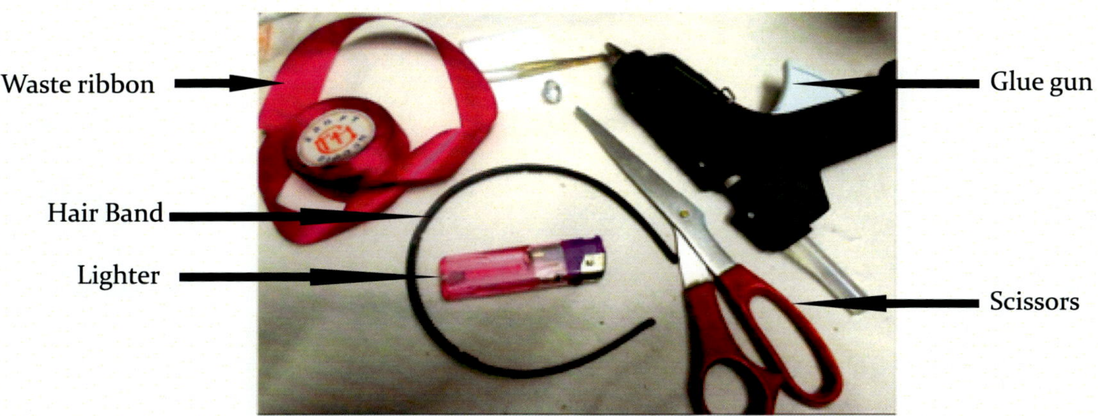

Fig. 1 Material Required

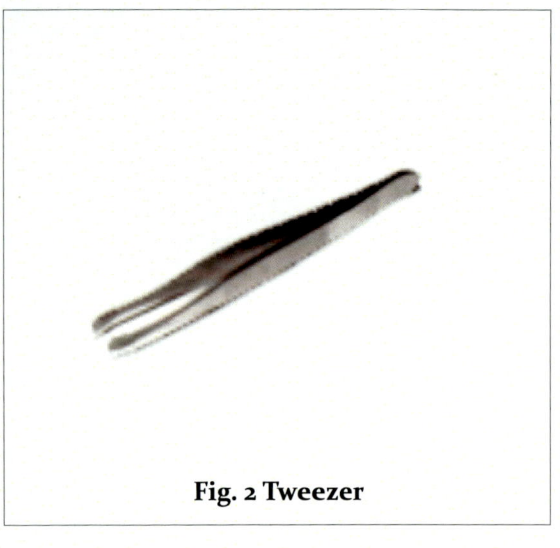

Fig. 2 Tweezer

Fig. 3 Crystal beads

PROCESS

Step :1

Fig. 1.1 Cut the ribbon into 1-2 inches in length.

Fig. 1.2 With a lighter burn the raw edges of ribbon so that they fuse and do not allow the ribbon fabric to ravel.

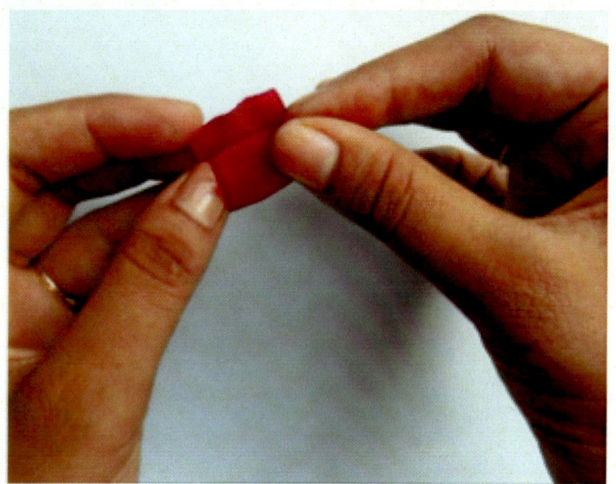

Fig. 1.3 Fold the ribbon as shown above.

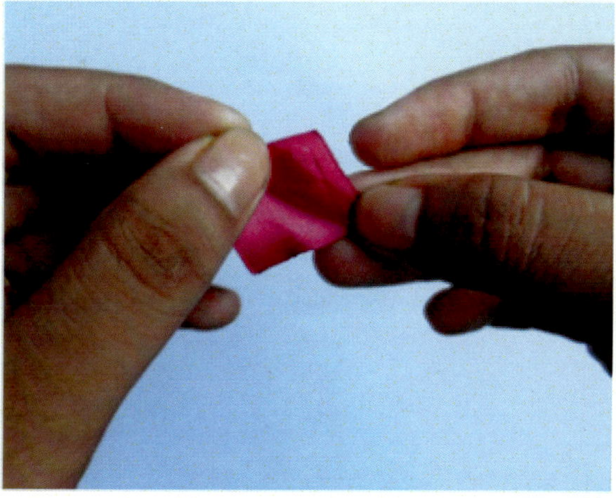

Fig. 1.4 Turn over one end of the ribbon in triangular shape. Follow this step on both sides.

Step :2

<u>Folding Steps</u>

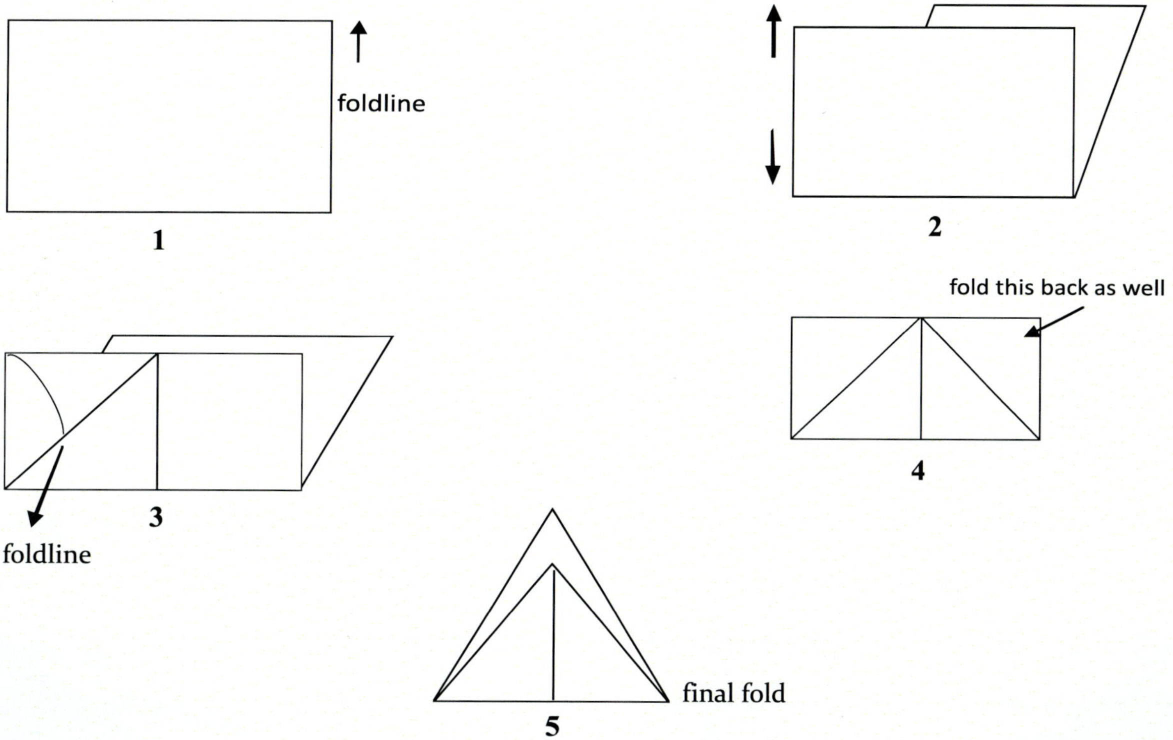

foldline

1

2

fold this back as well

3

foldline

4

final fold

5

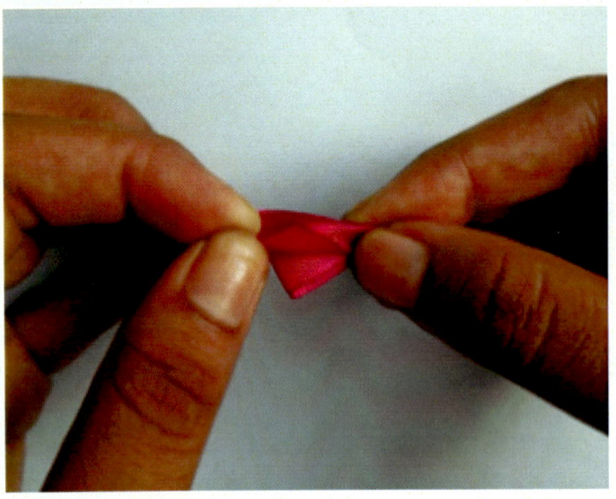

Fig. 2.1 Now fold the smaller triangle from both sides.

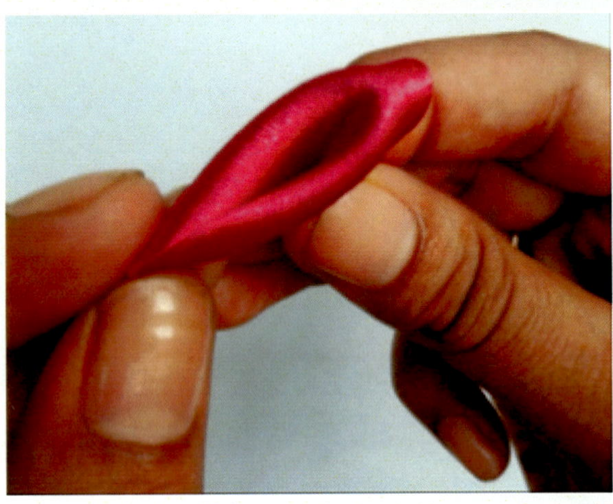

Fig. 2.2 This will form a "petal like" shape at the reverse side.

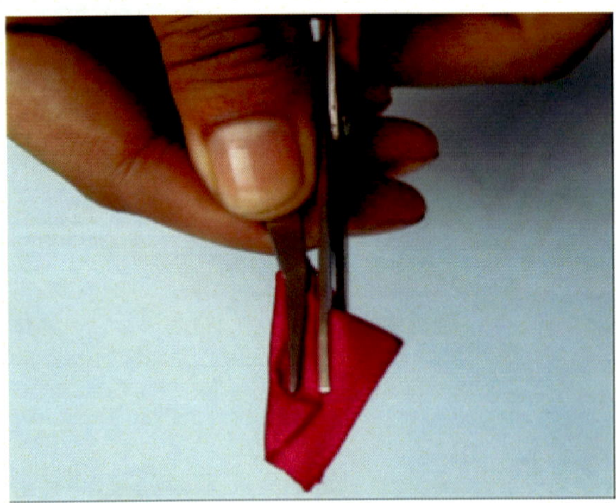

Fig. 2.3 Hold this shape properly between the tweezers and cut the extra ribbon.

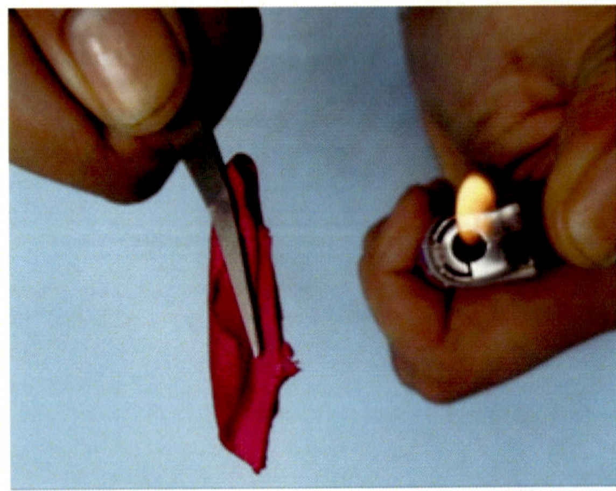

Fig. 2.4 With the flame of the lighter fuse the fabric together.

Fig. 2.5 The petal is ready. Make as many petals as you want, depending on the size of the flower to be made. Here I have made 24 petals.

Step :3

Fig. 3.1 Cut a clean circular shape of 1" diameter in tetron material.

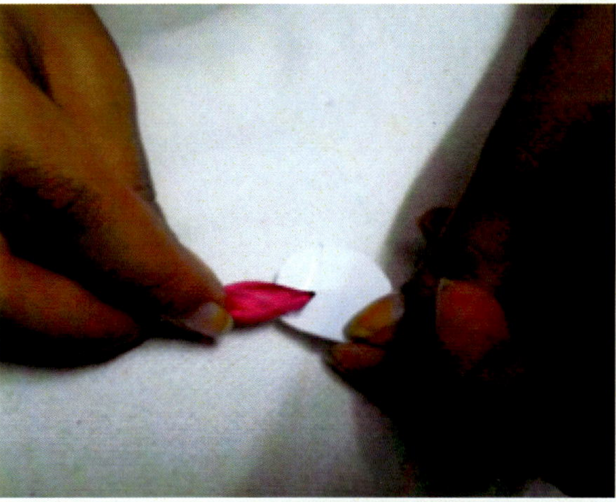

Fig. 3.2 Stick the petals one by one on the tetron circular piece with the help of glue gun or feviquick.

Fig. 3.3 The petals need to be fixed close to each other in order to create a beautiful flower.

Fig. 3.4 Continue sticking the petals.

Fig. 3.5 When the flower is made put some feviquick in the centre and place the crystal bead in the centre.

Fig. 3.6 Press firmly on the bead so that it fixes properly.

Fig. 3.7 Drop some glue/feviquick on the back side of the flower so as to stick it to the band you have bought from the market.

Fig. 3.8 Place the band on the feviquick and hold it firmly so that it sticks properly.

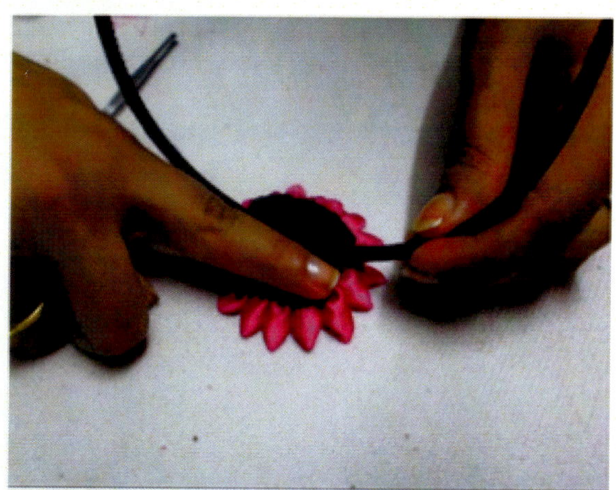

Fig. 3.9 Stick a piece of circular cotton fabric of diameter 1 inch over the white circular tetron piece so that it gives a cleaner look on the reverse side.

Fig. 3.10 Your headband is ready. You can experiment with double shades of petals or you can use a bunch of flowers. Have a lot of fun in this creative way!!

PRODUCT-3 : Flower Choker Necklace

Fig. 1

Material Required

1. Black satin (24cm length and 2"in width)
2. Red satin (24cm in length and 2"in width)
3. Black and Red beads (10 mm thickness)
4. Steel hook and eye
5. Cord for necklace (1-2mm thick polyester, nylon, elastic, cotton, etc.)
6. White thread for stitching
7. A 7 number hand sewing needle
8. Cigarette lighter (smoking kills!! ☹)
9. A 12" scale
10. A pair of 4 inch blade scissors

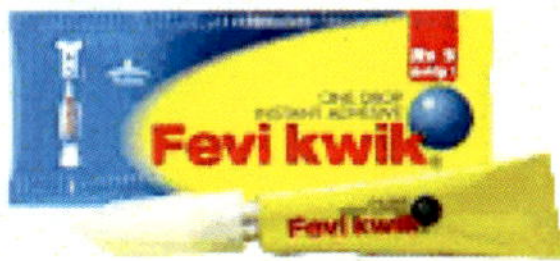

Glue

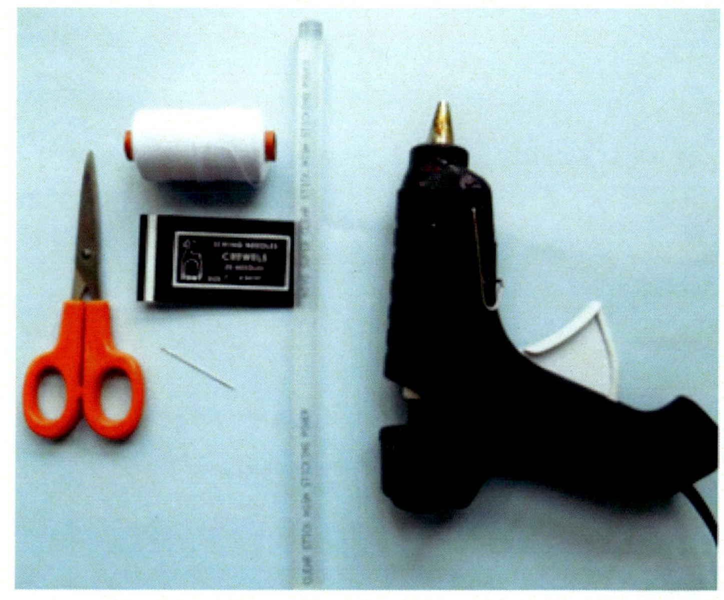

Scissor, thread, needle, glue gun and glue stick

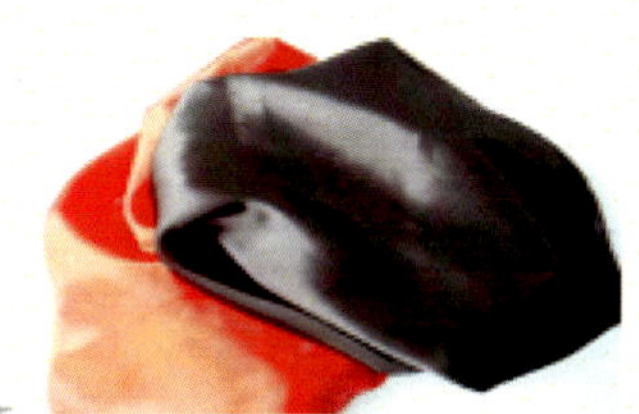

Satin fabric

Black and Red beads

Hook and eye

Polyester chord

Fig. 2

PROCESS

Step-1

Fig. 1.1 Cut the satin fabric in strips 21 inches long and 3 inches wide.

Step-2

Fig. 2.1 Finely finish the sides of the satin strip and fuse the raw edges with the help of a flame of the lighter

Step 3

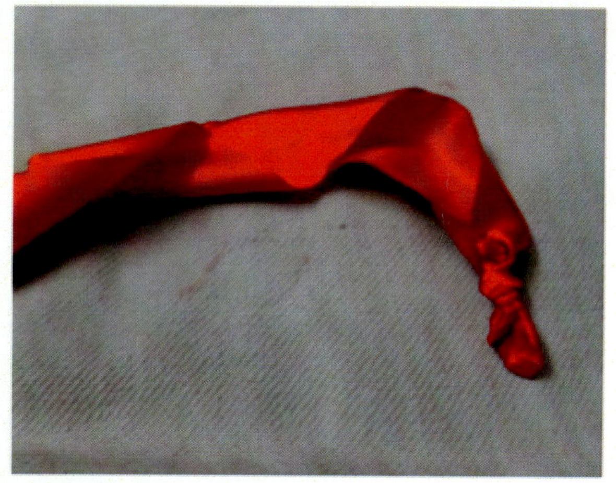

Fig. 3.1 Tie a knot at one end of the fabric strip winding it around the fabric itself as shown above.

Fig. 3.2 Twist the strip and continue twisting as you can see in the above picture.

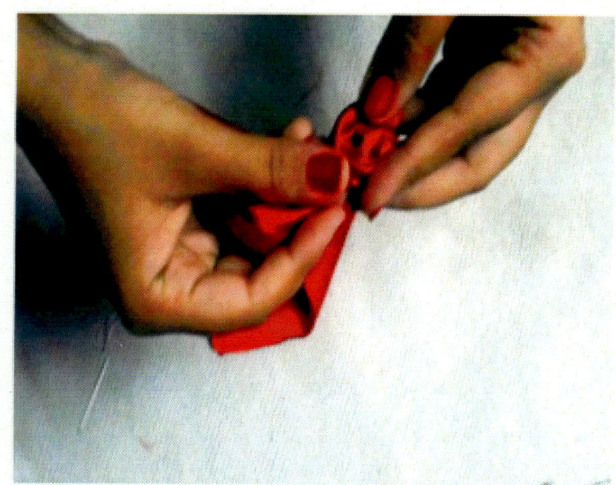

Fig. 3.3 Keep rolling and fixing it by sewing it with thread as you go along.

Fig. 3.4 You have just rolled your satin fabric in a flower shape.

Step 4

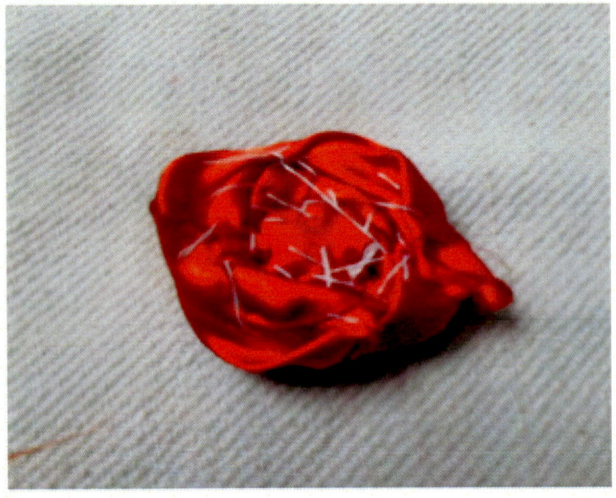

Fig. 4.1 The back side of the flower looks like the picture above.

Fig. 4.2 The front side of the flower looks like the picture above.

Make a total of three flowers; two red and one black.

Step-5

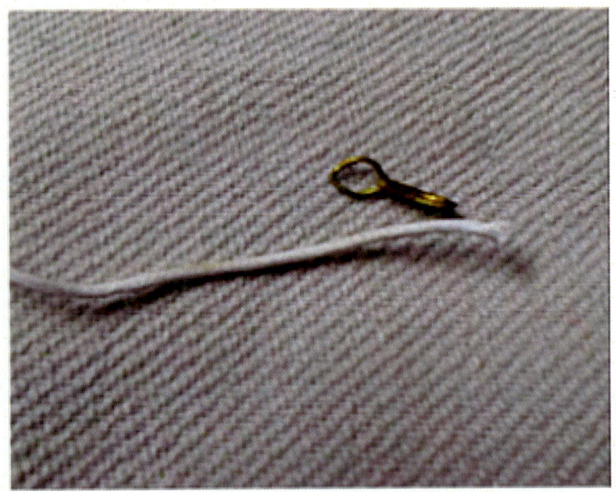

Fig. 5.1 Cut 25cms length from the 2mm chord.

Fig. 5.2 Firmly fix a hook at one end of this cord by tying.

The cord can be nylon, elasticated thread or any firm thread you think will be suitable.

Step-6

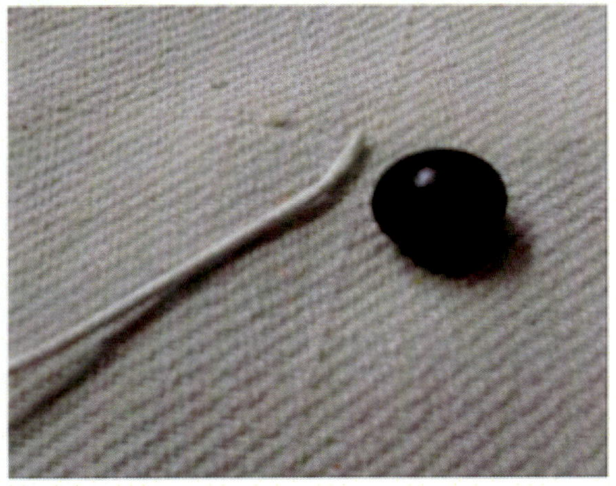

Fig. 6.1 Thread the black and red beads alternately or in any sequence you desire.

Fig. 6.2 Put the beads till the half of the final size (about 12cms) in length. The ready size in this is 24 cm. Half size is 12cm.

Step-7

Fig. 7.1 Hold the flowers together and firmly glue or stitch them together in a row with a needle and thread.

Fig. 7.2 Do not let the stitching thread show. A white thread is used in the above pictures only for demonstration purpose.

Fig. 7.3 Fix the three flowers firmly to the cord one by one with the help of thread and needle.

Step-8

Fig. 8.1 Continue to add more red and black beads in the same combination as the first half you made earlier.

Step-9

Fig. 9.1 Finally thread the closing hook and fix it firmly. Your flower Choker Necklace is ready to use. You can show it off to friends!!

Product 4: A Graceful Bow Clip

Fig. 1

__Material Required__

1. A waste piece of net fabric
2. 4" blade scissors
3. Thread
4. Hair Pin
5. Glue gun /Permanent Glue (Feviquick)

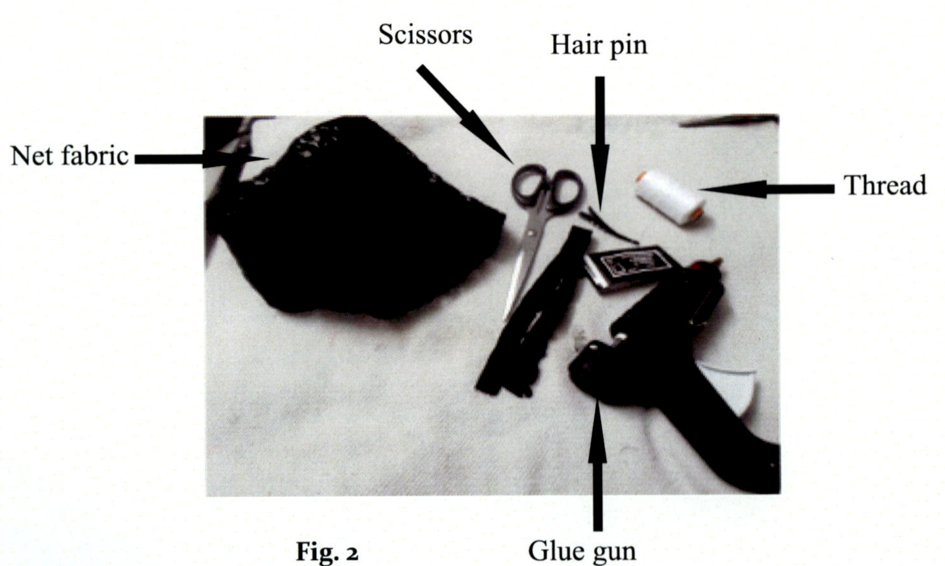

Fig. 2

PROCESS

Step 1

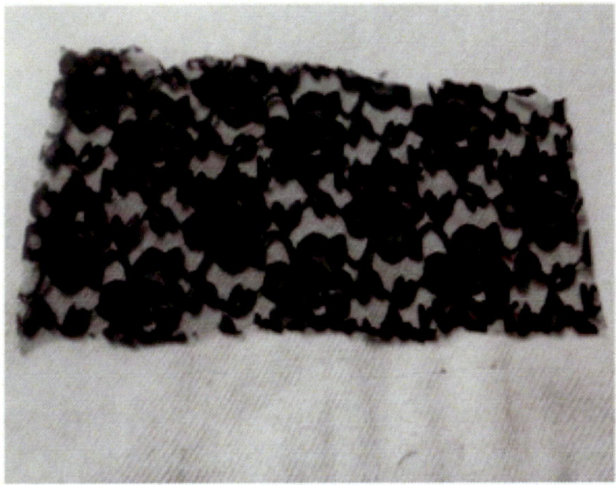

Fig. 1 .1 Cut the net fabric in 12" by 5" size

Step 2

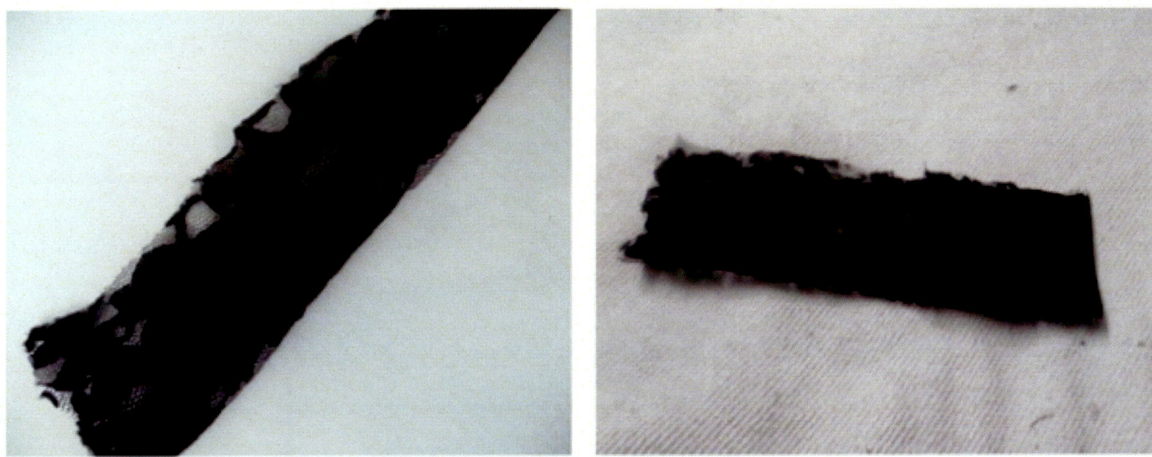

Fig. 2 .1 Fold the fabric width wide so that it is 2".

Fig. 2 .2 Stitch the edges in order to hold the sides 2.5" wide together.

Step 3

Fig. 3.1 Twist and bring the seam to the middle as shown in the figure.

Step 4

Fig. 4.1 Now fold this step as shown in the picture; placing one fold over the other, till the entire length gets folded to 2".

Step 5

Fig. 5.1 Hold the fabric from the center as shown above. Stitch and fix this shape.

Step 6

Fig. 6.1 Cut a small strip of same net fabric of 10 inches length and ½ inch width. Stitch and finish it as you did earlier. Use this strip to tie it in the center to hold the folded net fabric in a bow shape.

Step 7

Fig. 7.1 Sew the strip to it firmly on the reverse side.

Fig. 7.2 Lock the stitch a shown so that the strip stays is place.

I have used a white thread here to demonstrate. But the thread used should be the same color as that of the net fabric used.

Step 8

Fig. 8.1 Apply feviquick to this (reverse)side of the bow with the help of a glue gun.

Step 9

Fig. 9.1 Place the Hair Pin on the glued surface and press it with hand to stick it. Apply feviquick to the back side of a hairpin as well. Press and hold it down till it is firmly fixed. The bow is ready to be used!!

Note: Pre heat the glue gun a few minutes before using it.

REFERENCES:

1. Reducing costs through waste management guide: the garment and household
2. Waste Minimization and Total Productivity Maintenance; Volume 3
3. Technical Manual: American association for testing &materials (ASTM), Vol.7.1&7.24. en.mimi.hu
4. Article on "waste management in garment industry" –Shubh Raha ,Manager PPC Department at Bhartiya International.
5. http://en.wikipedia.org/wiki/Waste_management textiles sector, june1997

Sumita Sikka
Email: sumitasikka@gmail.com
Assistant Professor
SD College, Sector 32 , Chandigarh

I take immense pleasure in sharing my journey undertaken to convert silk scrap into a beautiful bag. You can do it too by following these step -by-step instructions which shall guide you through the process. The main interest element of the bag is surface enhancement by "layering technique" done on both the front and backsides of the bag.

Product 5: Draw Sling One Piece Bag

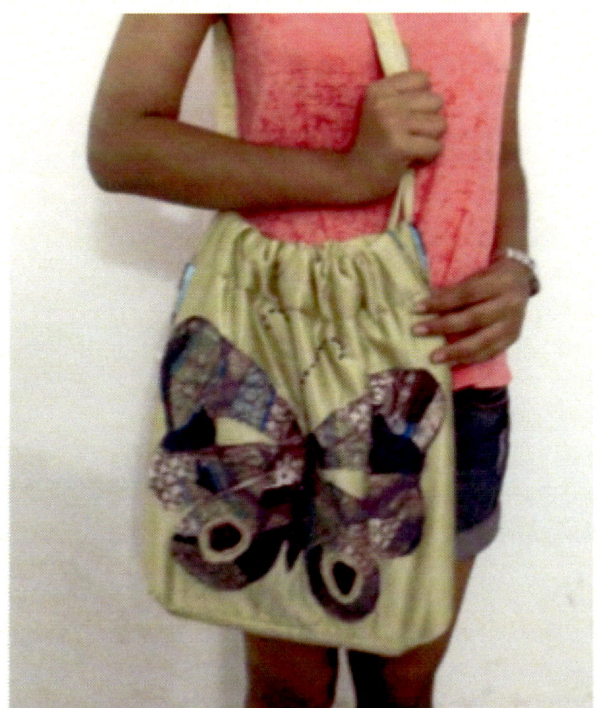

Fig. 1 Bag Front

Fig. 2 Bag Back

Material Required

1. Cotton silk for outermost shell: 1 meter (This may also be prepared by joining several small pieces of same hue)
2. Fleece for inter lining: ½ meter (1/4" thick foam or discarded layers of muslin may also be used)
3. Satin for lining: 1 meter
4. Paper fuse: 25" × 17" piece and 44" × 5' piece
5. Sewing threads: Blue, purple and golden color
6. Silk scrap: Silk pieces of any dimension more than 2"
7. Single needle lockstitch machine
8. Iron and Ironing Board

All textile material should be discarded, waste material.

PROCESS

Step-1

a) To make a bag 15" × 11" with a 4" wide base, cut a rectangle in cotton silk fabric and fleece piece measuring of 40" × 17".

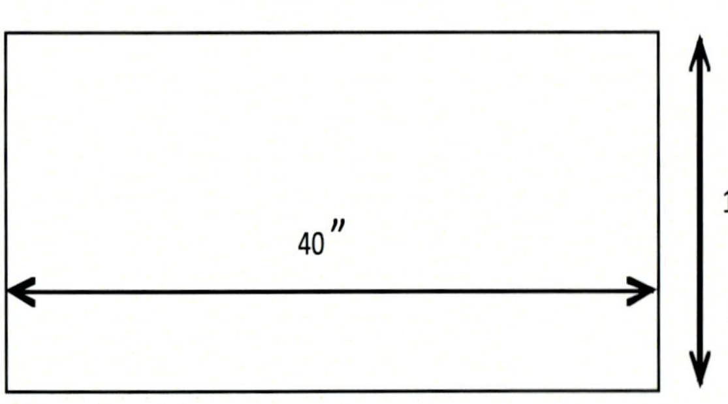

Fig. 1.1 Dimensions of cotton silk piece and fleece

Fig. 1.2 Dimensions of bag

b) Choose a colour scheme you want to work in. You can select any suitable picture from a magazine for this. Use it as reference to select various colour of scrap fabric. Try to stay as close as possible to the selected colour scheme.

c) Select a picture you would like to use as a motif on your bag. Here I have chosen a butterfly

Fig. 1.3 Colour inspiration

Fig. 1.4 Motif inspiration

Step 2

a) Select cuttings of equal weight fabric as per colour scheme chosen and iron them out.

b) Here random shapes of Silk and Brocade scrap have been chosen to create a butterfly motif on the front side of the bag. Straight strips of similar fabric have been chosen for the back side of the bag.

Fig. 2.1 Silk Scrap

Fig. 2.2 Silk Scrap

Step 3

a) To create a motif ,take a paper fuse of 25" × 17". Cut silk scrap in small pieces (approximately 2"-4") and place these pieces randomly on the paper fuse taking care to distribute colour of these scrap pieces as per your colour inspiration taken above.

Fig. 3.1 Place the pieces of scrap randomly on paper fuse.

Fig. 3.2 Fuse the fabric scrap to paper fuse using a hot iron.

b) Place any cotton fabric over the scrap arranged on paper fuse and iron at maximum temperature recommended for cotton fabric.

Step 4

a) Stitch randomly on all the pieces in order to secure them to the paper fuse. Make sure all the pieces are secured properly in their position

Fig. 4.1 Stitch on silk pieces to bind their ends on paper fuse

Fig. 4.2 Silk pieces secured properly in their position.

b) Place 12" × 17" of this prepared layer of silk scrap over a layer of cotton silk fabric and fleece of size 40" × 17" cut earlier in step 1 a). Save the rest of silk scrap Spiece for the back.(You can use foam or 3 to 4 layers of muslin fabric instead of fleece).

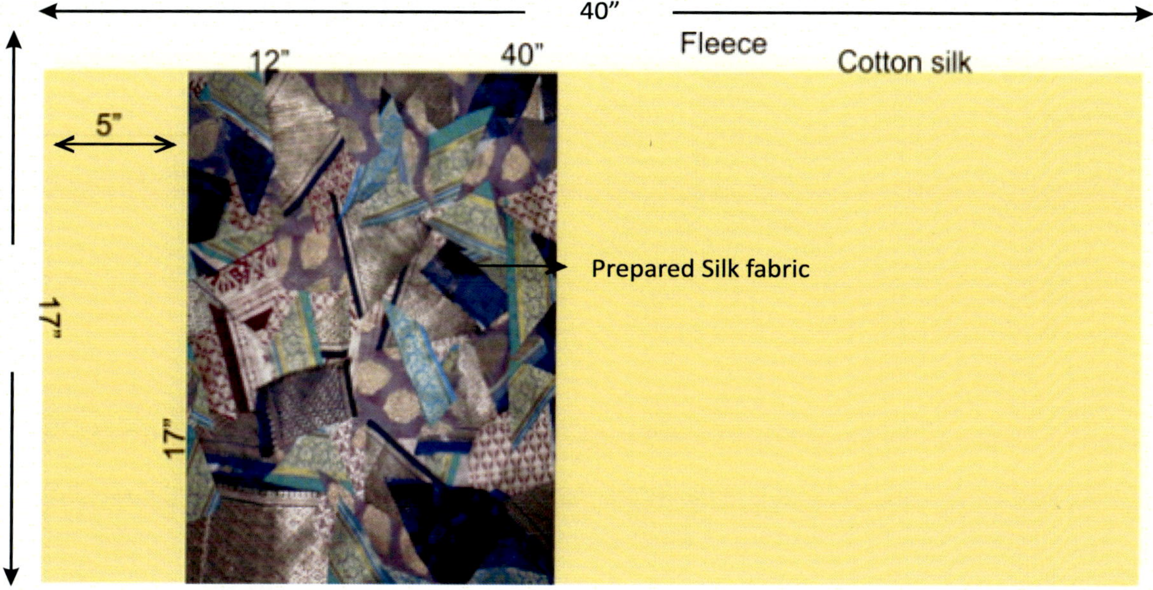

Fig. 4.3 Placement of silk scrap fabric, cotton silk and fleece

c) Now you need to draw the butterfly motif on the back fleece layer, exactly on the reverse of silk scrap piece.

d) For placement of the motif measure down 5" from the top edge of the layer. Turn back all the three layers together.

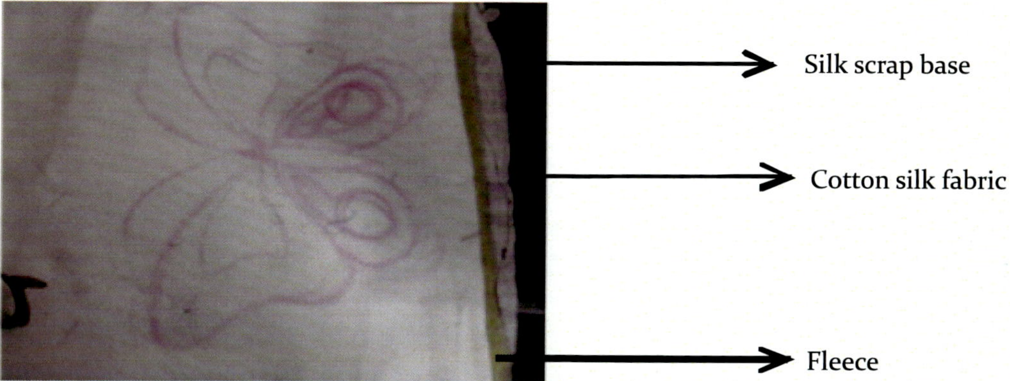

Fig. 4.4 Drawing of motif on backside of fleece

e) Draw a selected motif within 15" x 12" are on the back layer, which is the fleece layer. You can use any marker pen or chalk to draw the motif.

f) Place or sandwich some more silk scrap pieces in selected solid colors just under this drawn motif. These will enhance the shades of butterfly wings later on.

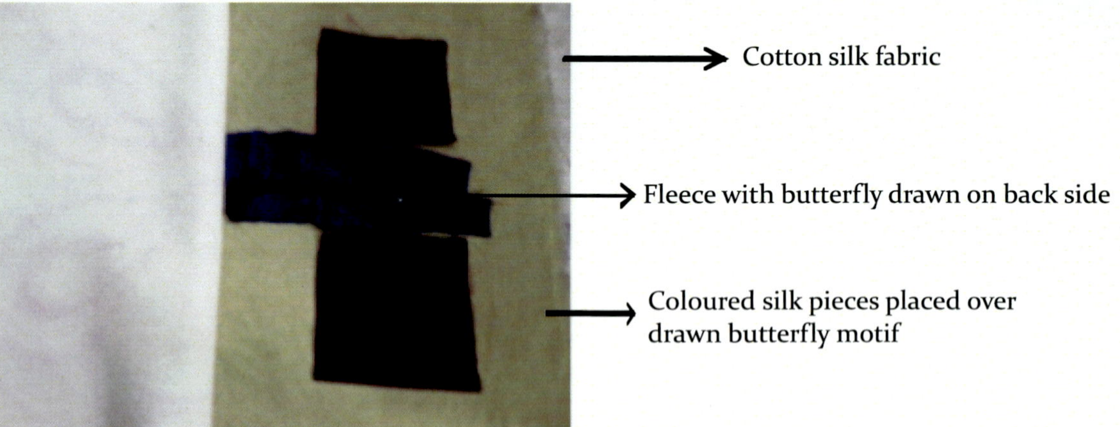

Fig. 4.5 Sandwiching solid coloured silk piece

Step 5

a) Holding the backside (fleece side) facing you, machine stitch around all the sketch lines of the motif. Keep the SPI (stitches per inch) up to 20-22 stitches

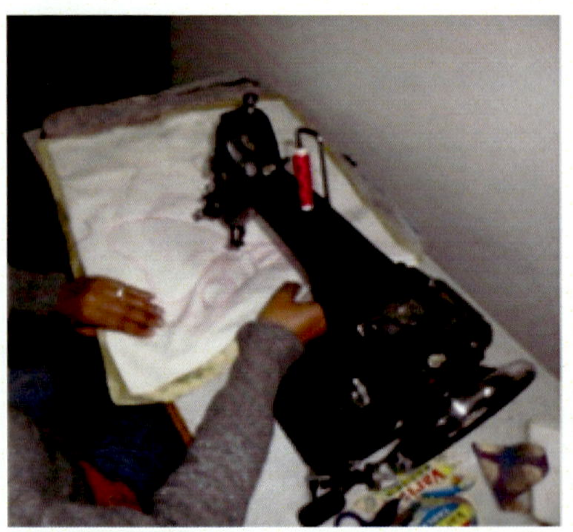

Fig. 5. 1 Machine stitch on the sketch lines of motif

Fig. 5. 2 Stitched motif

b) Turn over to the front and cut the fabric around the outline of the motif. Also snip through the fabric, inside the motif, to make the inside layers of fabric pieces visible. This gives a pretty 3D effect to the butterfly artwork. The front of the bag is ready.

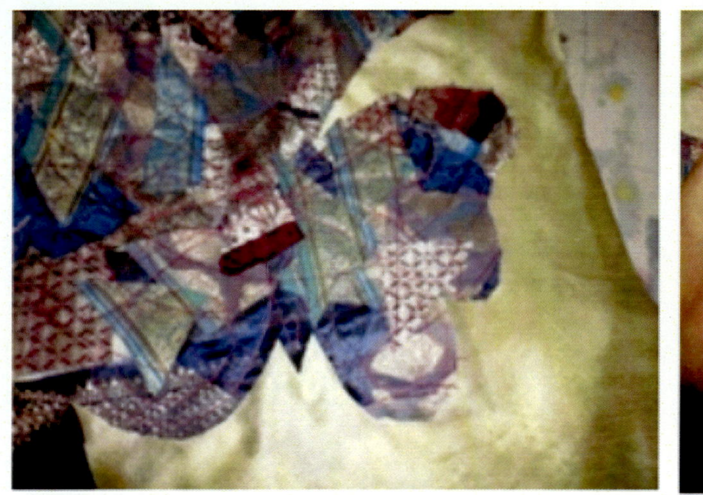

Fig. 5. 3

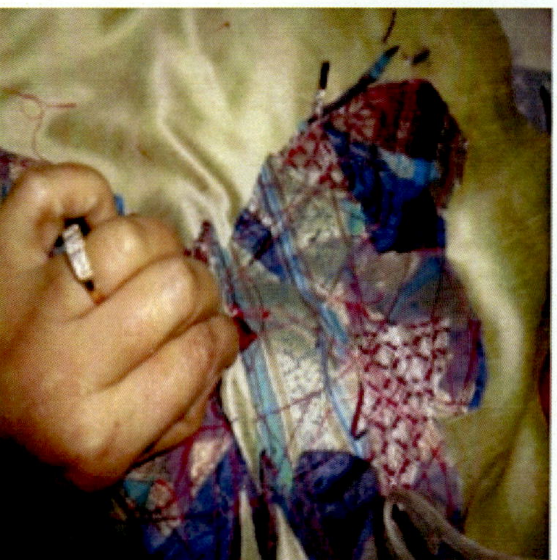

Fig. 5. 4

Step 6

Now prepare the back side of the bag by stitching strips of silk scrap and plain fabric.

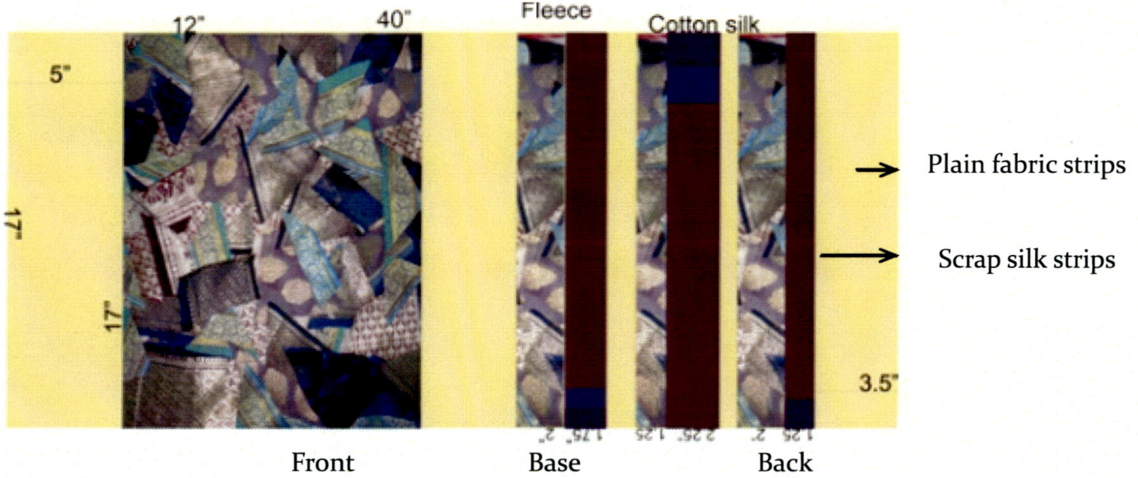

Fig. 6.1 Complete Outer shell

a) Use the silk scrap pieces you saved in step 4b) for cutting strips for the back side of the bag.
b) Cut width of these horizontal strips are of 2.5" and 1.75".
c) On the cotton silk fabric mark straight lines at a distance of 2" and 1.25" and place the strips under the marked cotton silk fabric. Stitch on the marked lines.
d) Snip the cotton silk fabric carefully in parts and remove the fabrics to make the underneath fabric visible.
e) Prepare strips of plain silk fabric measuring 3", 2.5", and 1.5" width from plain silk pieces (if required give joints to complete 17" horizontal length).

f) Place them next to silk scrap strips after turning back 1/4" on either side of width to finish the edges.
g) Machine stitch throughout the horizontal length on upper and lower side.

The back of the bag is ready.

Fig. 6. 2 Marked lines on cotton silk face.

Fig. 6. 3 Folded strips of plain silk fabric

Step 7

Making the Base of the Bag

A bag needs to have a structured broad base for a smarter look and for better utility value.

a) Having completed step 6 g), we have a 40" long piece all set and ready.
b) Leaving 18" along the 40" length, pick and fold 1/2" thickness of fabric and machine stitch along the width as shown in Fig. 7.1
c) 4" away from this stitch, fold 1/2" thickness of fabric again and machine stitch in the same manner.
d) These stitching lines will make front and back look separate and also provide a structured base to the bag.

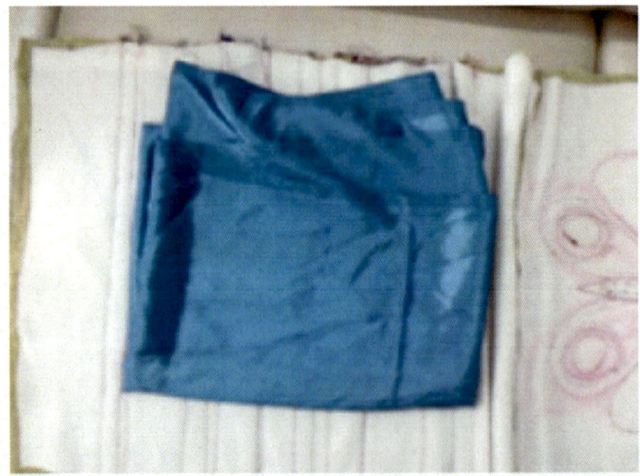

Fig. 7. 1 Stitching line for creating base

Step 8

Finishing up:

The bag is finished with satin lining and patch pockets of contrasting colour.

a) Cut a piece of satin fabric for lining, measuring 40" × 17". Also cut two pockets of sizes 10" × 14" and 13" × 14" each.These sizes include seam allowances.

b) Fold both the pieces of pocket into half along the length (14" side). The fold line makes the top of the pocket.

c) Pockets of 9" × 6" are prepared by ironing and folding back seam allowances. Extra width of one pocket, which is 13", is used for making box pleat.

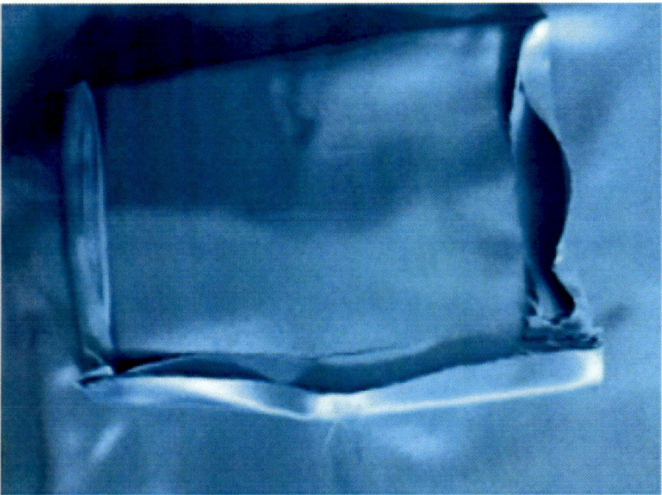

Fig. 8.1 Pocket 1

Fold back seam allowances from three sides

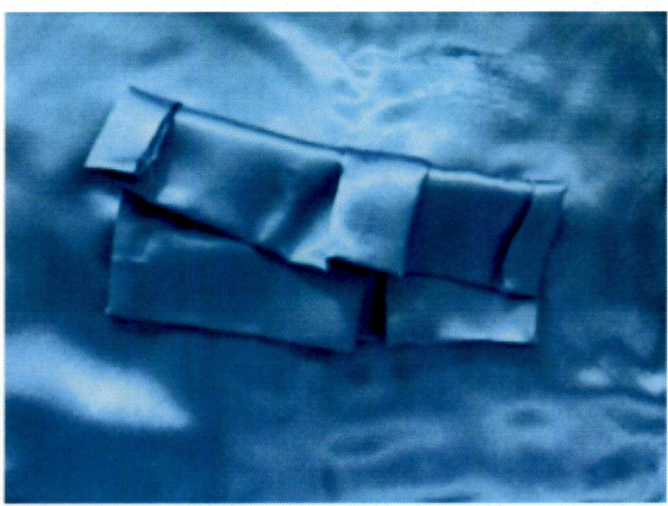

Fig. 8.2 Pocket 2

Fold back seam allowances from three sides and form a box pleat of 1 ½" for partition in pocket

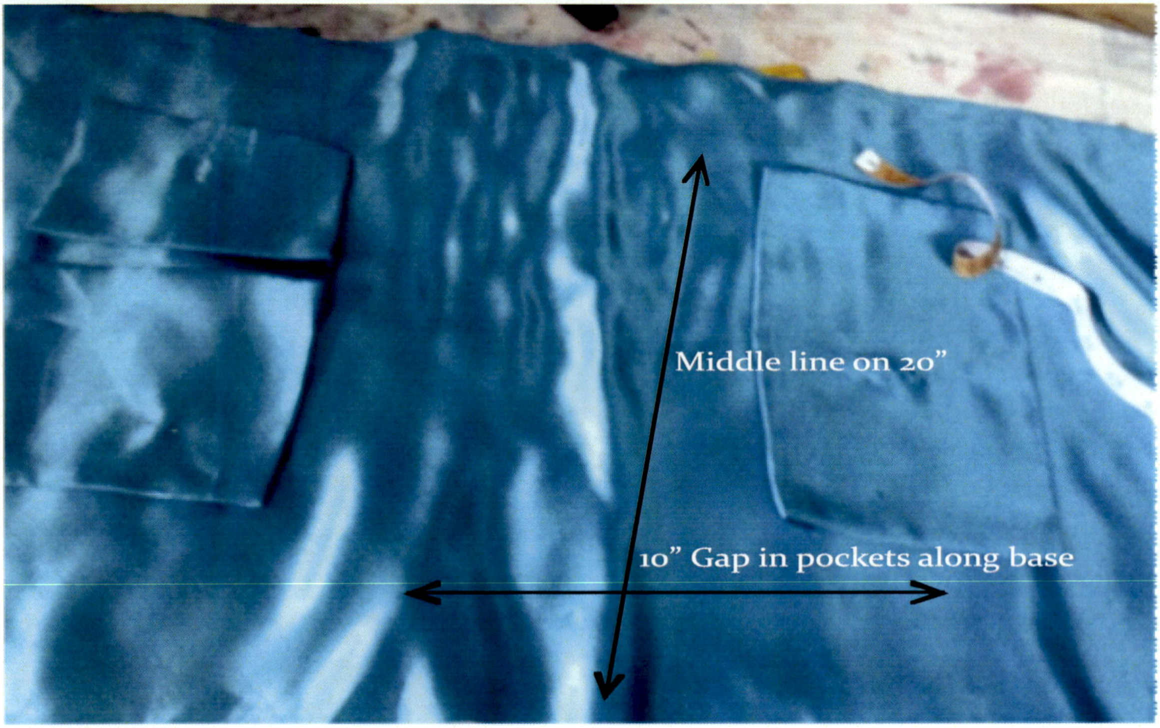

Fig. 8.3 Placement of pockets on satin lining.

d) Pin the pockets 5" from the middle line. This creates a 10" gap between the pockets.

Fig. 8.4

e) Stitch on the top edge of pocket to hold box pleat.

f) Stitch pocket on satin base from three sides and in the middle for partition.

Step 9

Attaching the Satin Lining

a) Place face of satin fabric over face of finished outer shell i.e. two rectangles of 40"x 17" together.
b) Stitch along length on all the four sides for 6" length and bag out.
c) Picture below is taken after bagging out face of both the fabrics. This step is required to have finished edges in the casing area.

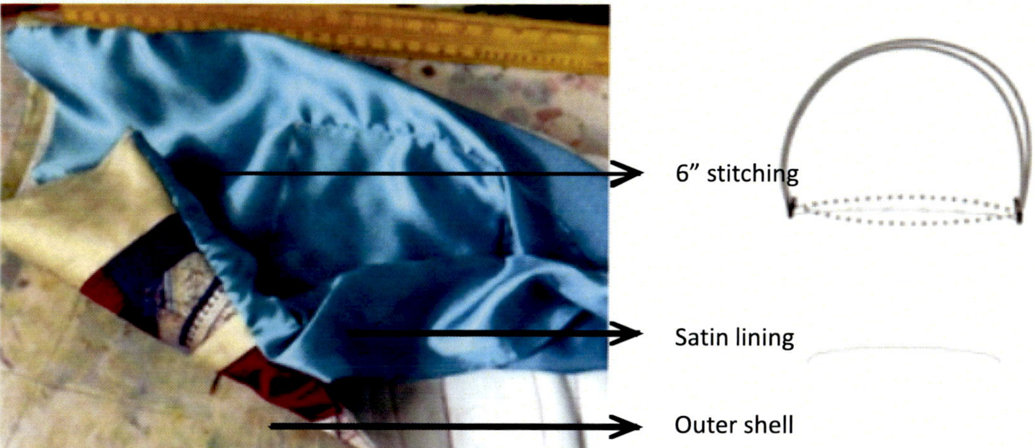

Fig. 9.1 Finishing of casing and opening

d) Holding right sides of the shell facing together stitch all around leaving 6" from top for making ready casing of 2.75". Do the same with the satin lining.

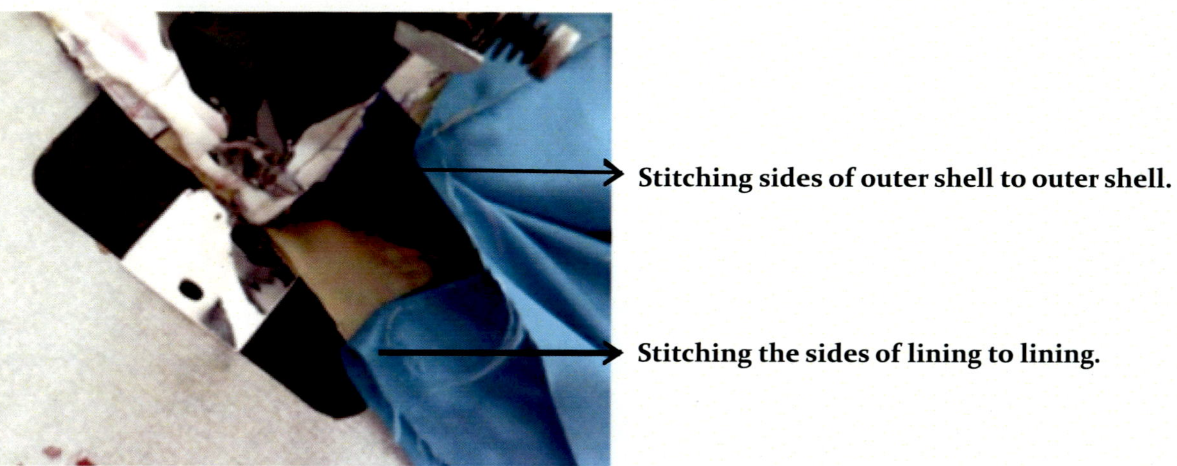

Fig. 9.2

e) For the base push out and shape a corner and stitch in a direction perpendicular to the middle of the base.
f) Cut the corners to reduce the bulk.

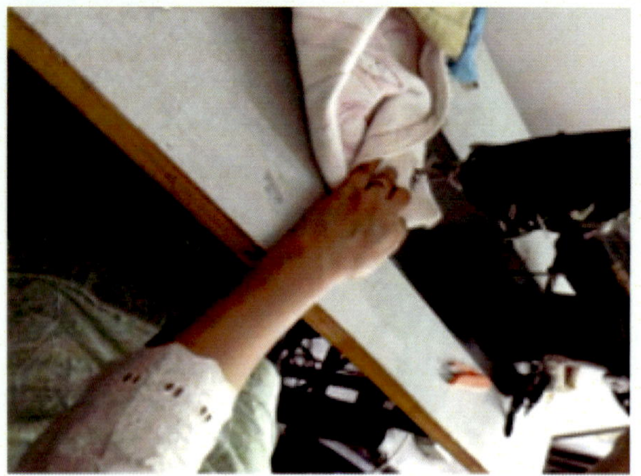

Fig. 9.3 Stitching for making base

Fig. 9. 4 Cutting the corners

g) The top is folded back by ½" and then 2 ½"for finished casing.

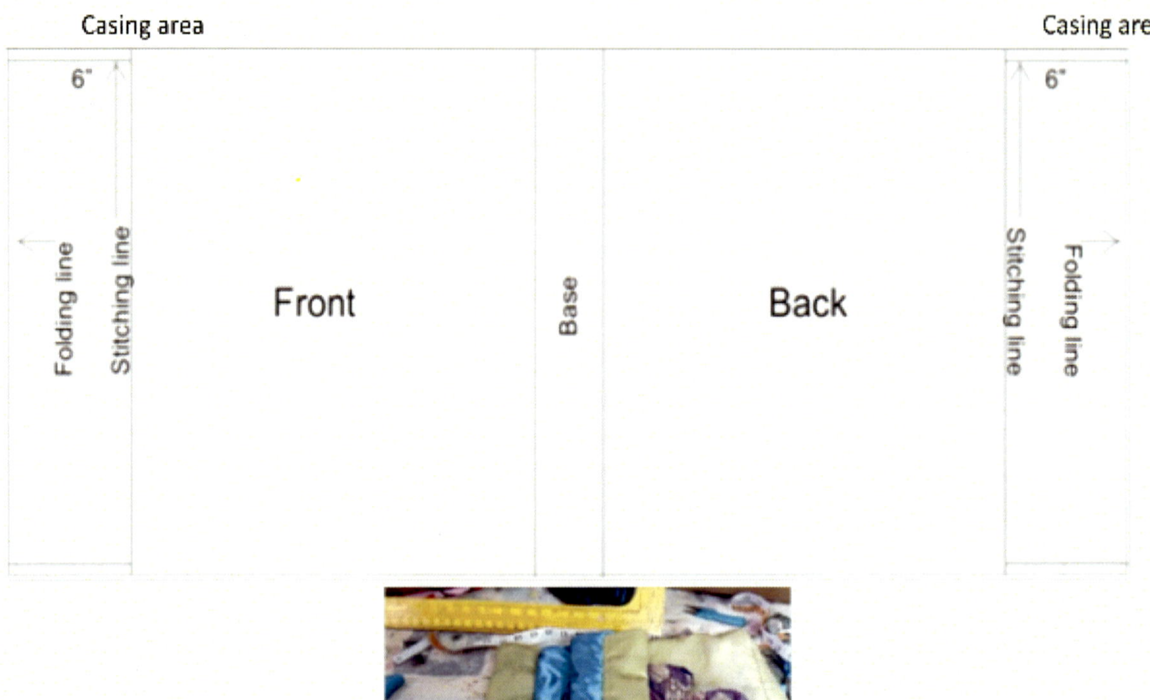

Fig. 9.5 Finished casing area

Step 10

Preparing sling of the bag

44"
2.5" Fold lines

a) Prepare two slings of size 44" × 2.5". To add strength to slings, paper fuse both the strips with same size (i.e. 44"×2.5") paper fuse. Fold the sides and iron them for having finished width of ¾".

b) Stitch on one side as other side is on fold.

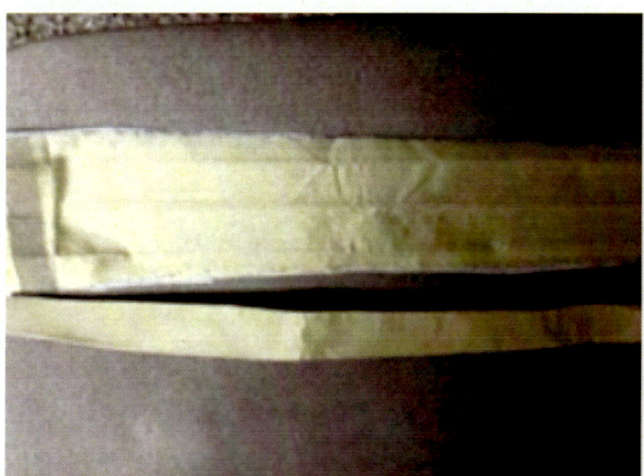

Fig. 10.1

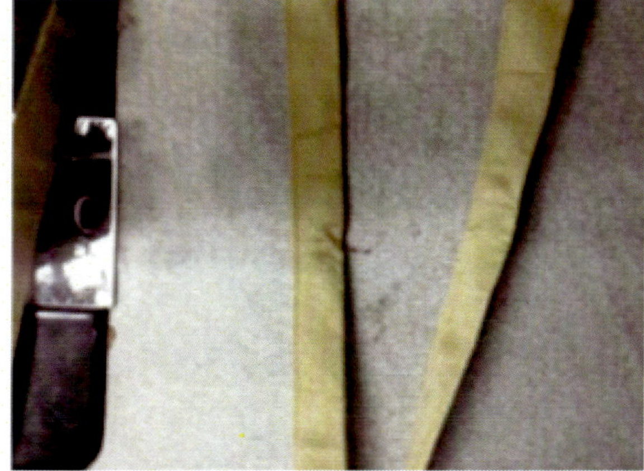

Fig. 10.2

a) Pass the string through casing of the bag and stitch together both ends to form a circle of 42" circumference.

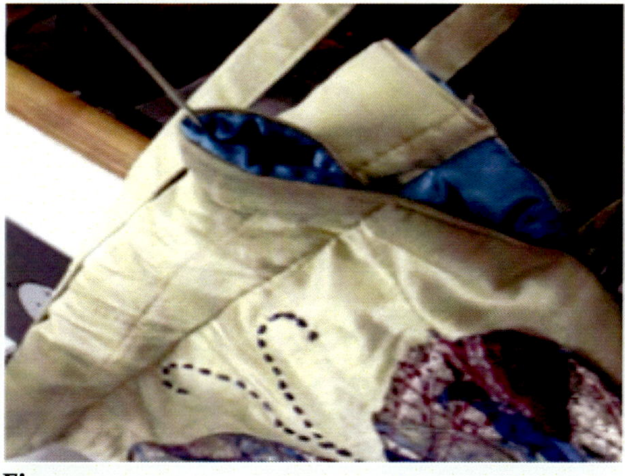

Fig. 10.3

Fig. 10.4

And Voila!! The bag is ready to be used.

The whole exercise was an enjoyable one and seeing the finished product was a great pleasure!!

Sugundha Saini
Email: sugundhas@yahoo.com
Assistant Professor
Amity University, Noida

Block Printing is a process of printing patterns on textiles, usually of linen, cotton or silk, by means of wooden blocks incised with patterns. It is a slow but very artistic way of value addition to textile products.

I have undertaken this creative skill project to sensitize and motivate my readers into doing simple projects of their own. Believe me, it is great fun and as you progress from one step to the next the excitement and sense of achievement grows without bounds.

Before I begin with the project I would like to advise to my friends, "Say No to Plastic Bags, Carry Your Own Pretty Shopping Bags".

Product 6: Block Printed Shopping Bag

Material Required

1. A discarded off white cotton shirt from my husband's wardrobe.
2. Scrapes from a waste sofa cover, sourced from home.
3. 1 Matching 16" zipper
4. Thread,
5. Fabric cutting scissor,
6. Tailor's measuring tape / a ruler 12"long,
7. Dress-makers pins,
8. Thread clipper,
9. Marking chalk.

Contd. on next page

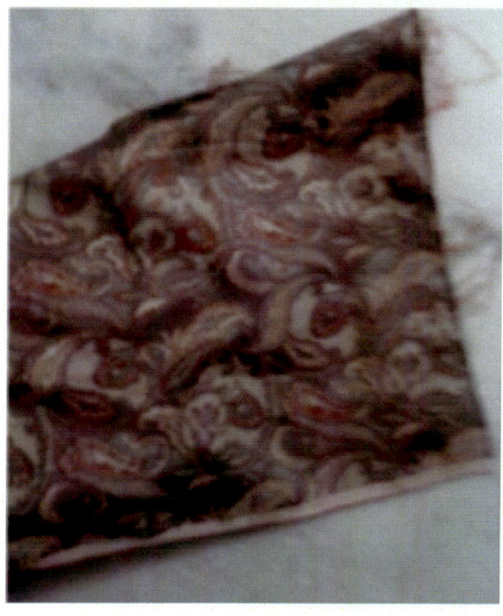

Fig. 2 Scraps from a waste Sofa Cover

Fig. 1 Discarded Shirt

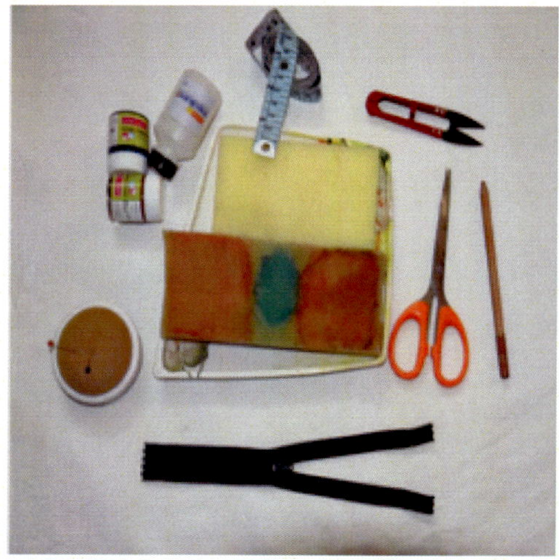

Fig. 3 A tray, a piece of sponge, fabric paints and medium to mix paint.

Contd. from pre page:

10. Padded table: Prepared by putting layers of folded 2-3 blankets.
11. Sponge - 8" x 4" x 1/2" thickness.
12. An enamel tray for mixing colour and placing the sponge pad
13. 2 Shades of fabric paint: rust and green in 15 ml packs of Fevicryl Brand.
14. Medium to mix paint: A 50ml bottle of Camel Brand.
15. Size 4 paint brush
16. 1" long safety pin
17. For my block print I chose six pre soaked wooden carved blocks of following sizes:-
    2" × 3" (Peacock Pattern) as A
    1.5 × 3" (Floral Pattern) as B
    2" × 2" (Camel Pattern) as C
    1.5" × 1" (Floral Buti) as D
    2" × 2" (Elephant Pattern) as E
    1.1" × 3" (Floral Border) as F.

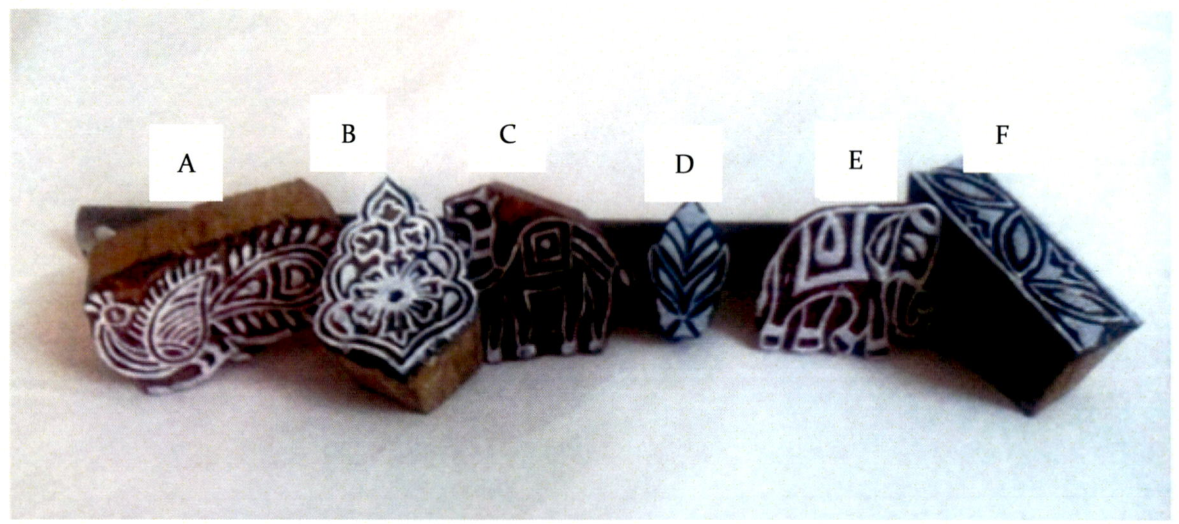

Note: New blocks are Pre-soaked in water over night so that they absorb minimal colour.

Step 1- Cutting

a) Cut plain white fabric, from the shirt, of size 32"× 16". This makes the main body of bag. If the fabric is not enough then you can take two pieces of 16" × 16" and join them. Here, I have used the front of a shirt, removed the buttons and ironed it out flat.

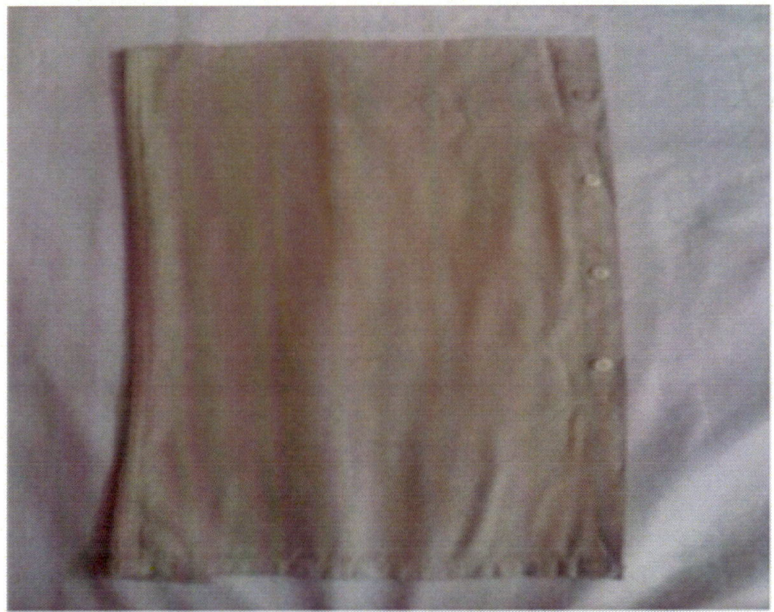

Fig. 1.1 32" × 16" fabric removed from the shirt front.

Fig. 1.2 Cut two strips of 16" × 5" from the scraps of waste sofa cover.

Fig. 1.3 Cut four pieces, 22" × 2.5" from sofa cover scrap to make the bag handle or sling.

Fig. 1.4 Two pieces of plain fabric from the shirt measuring 15"×2.5" for zip attachment.

Step 2: - Block printing the plain fabric that makes the main body of the bag.

a) Marking on the fabric: - The fabric is marked for printing as shown in the figure below, measuring 1" away from the sides with straight line for the border design, 1" away from top and bottom for Elephant and Camel motif and centre point for printing the Peacock motif.

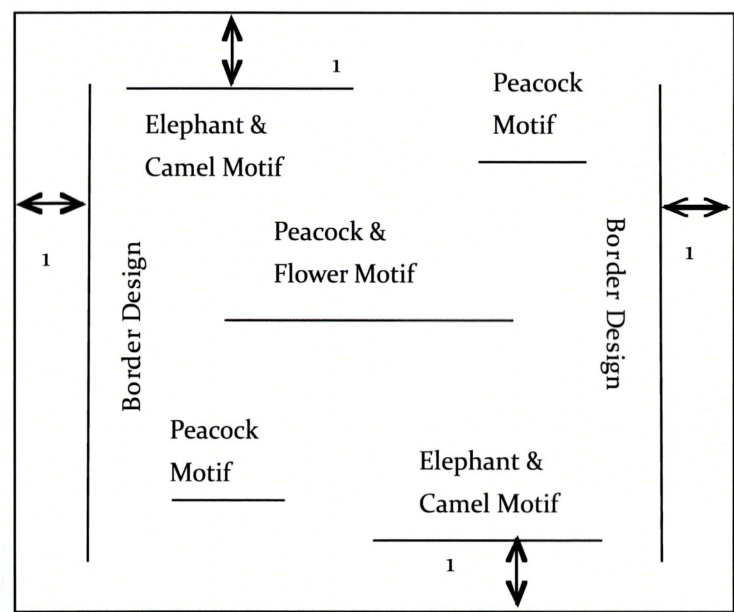

Fig. 2.1 Marking on the fabric

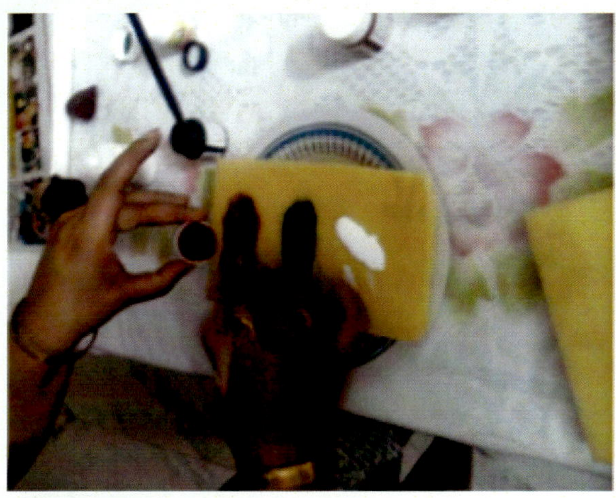

Fig. 2.2 Colour is mixed in the tray in the ratio of 2:1 i.e. 2 parts of colour and 1 part of medium. Then the colour is put on the sponge placed on the tray.

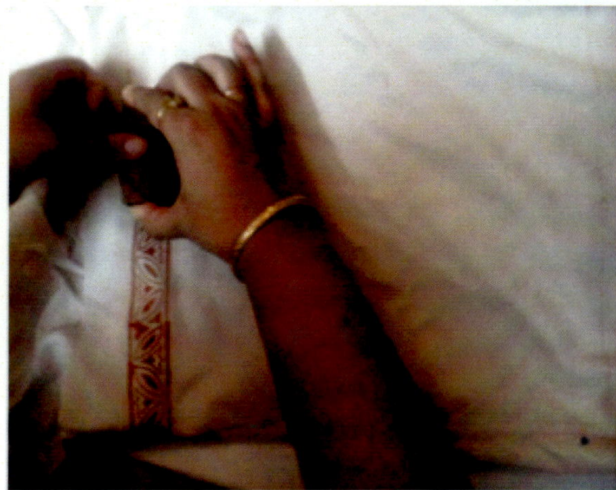

Fig. 2.3 Wooden blocks are selected according to the desired pattern and placed on the color soaked sponge. These blocks, holding the color are then pressed on to the fabric following the guidelines marked on the fabric. The blocks leave a fine desired print on the fabric.

Fig. 2.4 Blocks C, D and E are used by turns on top left and bottom right of the fabric.

Fig. 2.5 Then the block A is stamped in the centre and top right and bottom left.

Fig. 2.6 Peacock printed in the center

Fig. 2.7 Peacock printed on top right.

Fig. 2.8 Block B is stamped ½" on either side of centre block A.

Since, I had to repeat the printing with these blocks in combination, I fixed 3 of them to a metallic scale with a nail each one at a distance of 2" from the other. (Refer to the Fig. 2.9 and details on page no 59 on how to make this convenient tool)

Fig. 2.9 Convenient Tool

Step 3: Stitching the Bag

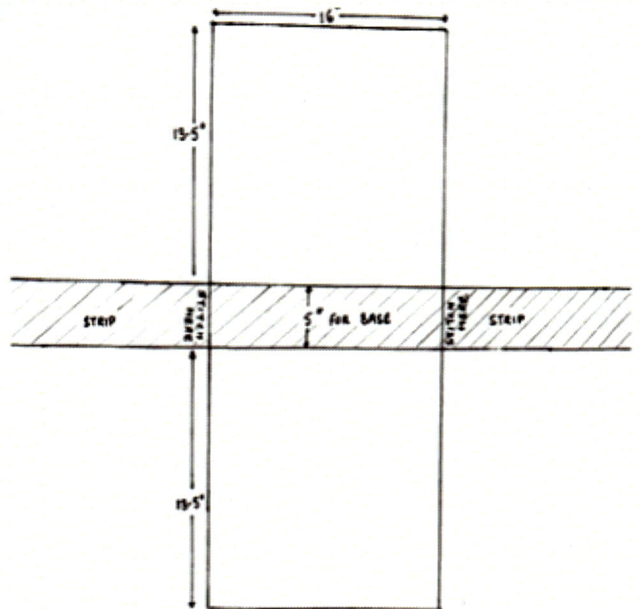

Fig. 3.1 For Making a Base: Attach the 5" width of the strips to the centre of the bag piece as shown

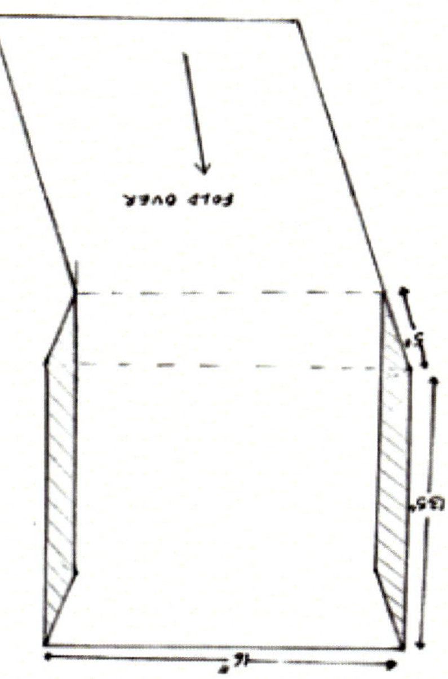

Fig. 3.2 Now attach the length of the strips as shown to the sides of the bag.

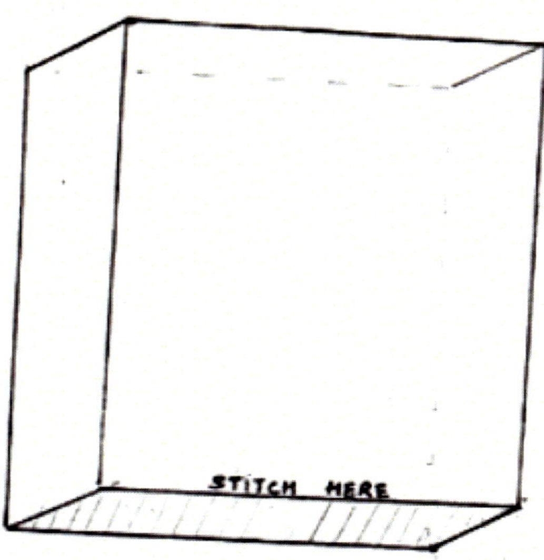

Fig. 3.3 Fold over the remaining fabric and stitch the sides.

Note: Use run and fell seam in all stitching or finish the raw edges of seams with a bias binding.

Step 4 - Preparing the Handles

a) Stitch the two 22.5" × 2.5" strips of fabric cut earlier along the two lengths. A tube gets formed. Turn these tubes inside out with the help of a safety pin. These will make two handles of the bag.

b) Find the midpoint of each strip and mark 4" either side of the midpoint. Roll over or fold this length and stitch this 8" distance to make a grip for handle.

Fig. 4.1 Ready Handle

Step 5 - Attaching the Handles

a) Sandwich the handle between zipper strip and bag piece

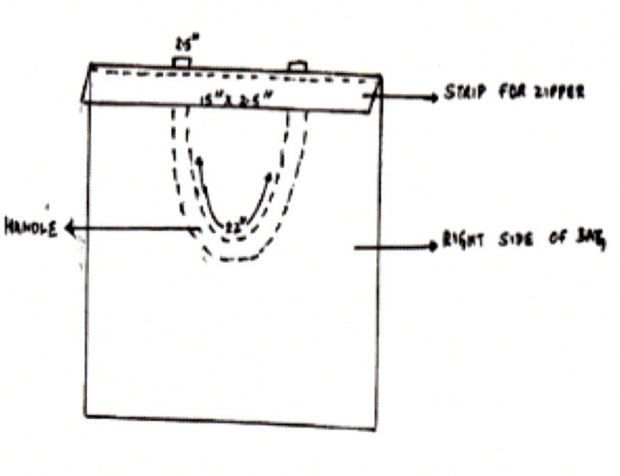

Fig. 5.1

Fig. 5.2

Step 6 - Attaching the Zipper*

a) Measure and mark the zipper's position on the seams.
b) Place the closed zipper on the two fabric pieces. Baste along both side of the zipper tape to hold it in place on the bag.
c) Turn the bag right side out. Start stitching at one end of the zipper tape and work around to the other end. In this manner stitch the zipper in place using the zipper foot if available. Keep the corners as neat and square as possible.
d) Avoid letting the stitches catch on the zipper teeth. Remove basting.

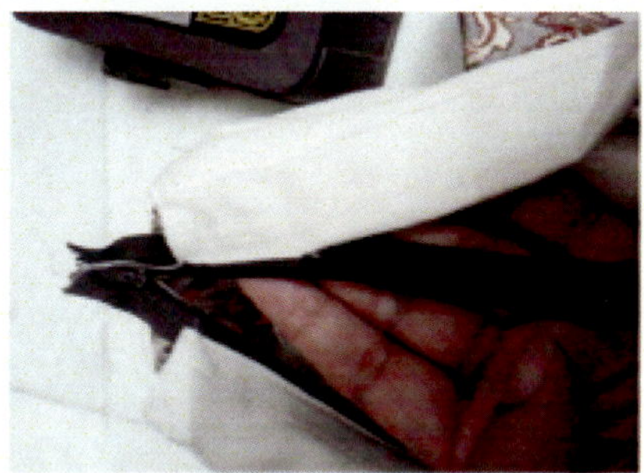

Fig. 6.1

Step 7 - Finishing Up

Cut off all hanging threads and any lint caught between the seams.
Your shopping bag is now ready!
Say no to plastic bags!! Bring in your purchases in your own shopping bag

Fig. 7.1

*Note: Zipper attachment taken from, The Ultimate Sewing Book, Maggi McCormick Gordon; pg. 104.

Product- 7 Bolster Cover

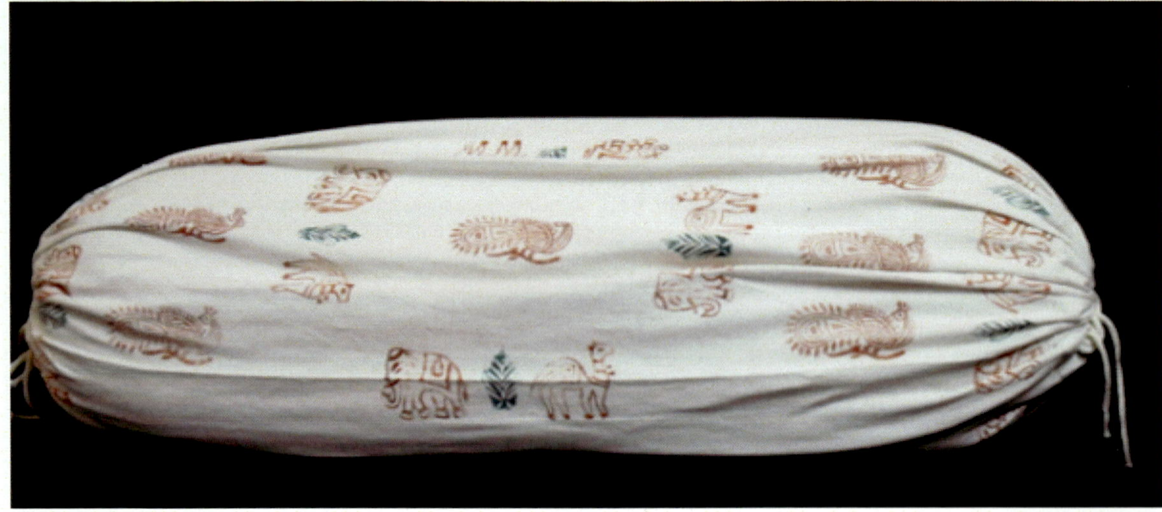

Fig. 1

Material Required

All material required for printing in Product 6, i.e. Printed Shopping Bag, can be used for making the Bolster Cover as well.

Also needed are:

1. Old Casement Bed Sheet
2. 2 Cords 40" long and 0.7mm thick.
3. Carved wooden blocks of chosen patterns. Same as taken for making Product 6 above.

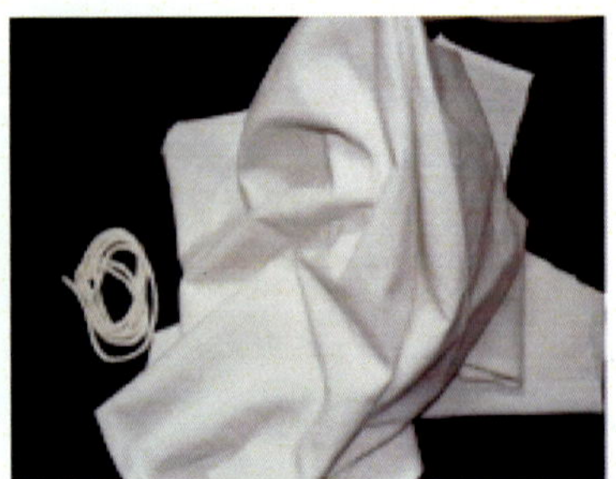

Fig. 2 Casement from discarded bed sheet and cords

Fig. 3 Blocks for printing

Step 1 - Block printing the plain fabric

a) Marking on the fabric: First, mark the area on fabric at a distance of 4" away from the sides and 4" away from center; make a square.

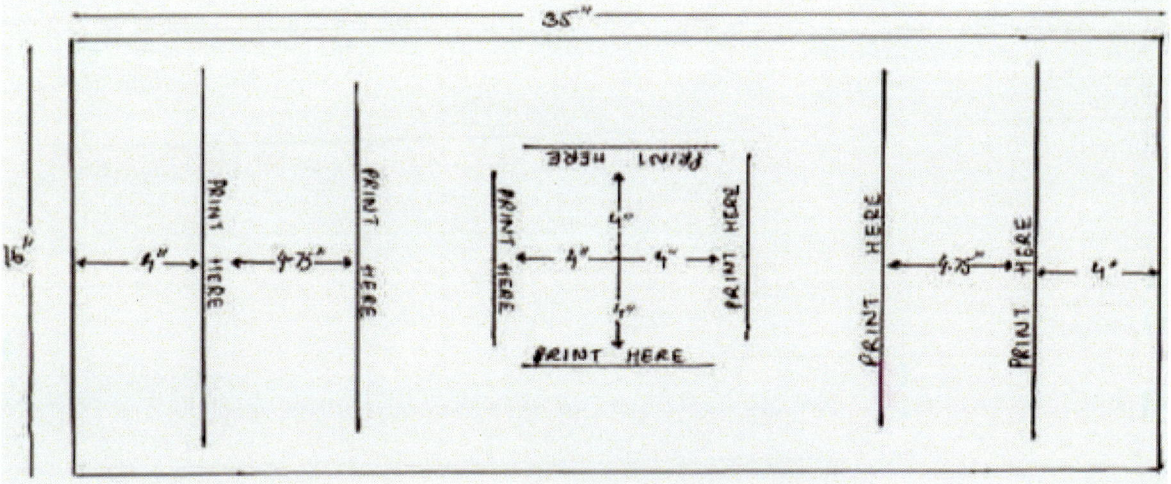

Fig. 1.1

b) Mix the colour in the ratio of 2:1 in a tray and pour it on the foam with the help of a brush.

c) Press the block on the foam to take up the colour and then print along the guide lines. Choose blocks of your liking.

d) Go on printing till the complete pattern is obtained.

Fig. 1.2 Central pattern.

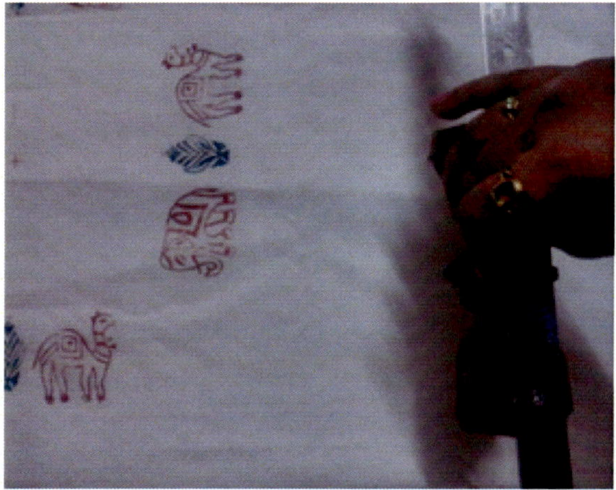

Fig. 1.3 Side pattern printing.

Step 2 - Stitching the Bolster Cover

a) Finish the 32" long edges first. Fold as shown in the figure to make a casing at the two ends for the draw string.

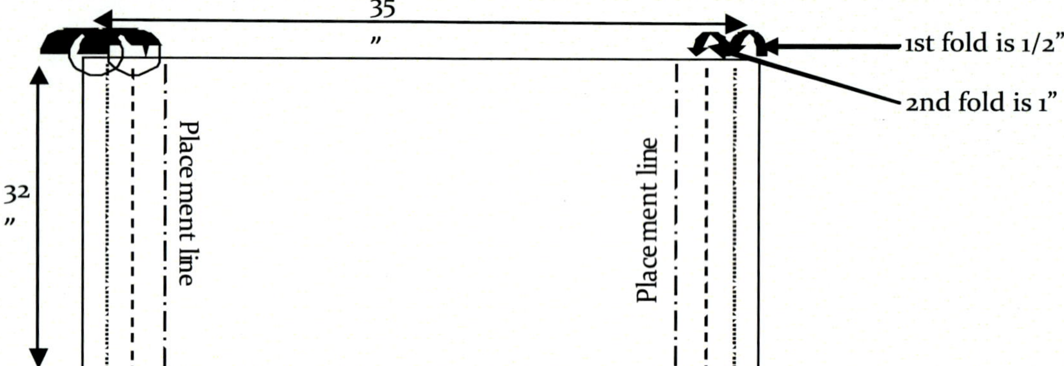

Fig. 2.1

b) Now bring together the two longer sides (35") and stitch to form a cylinder shape.

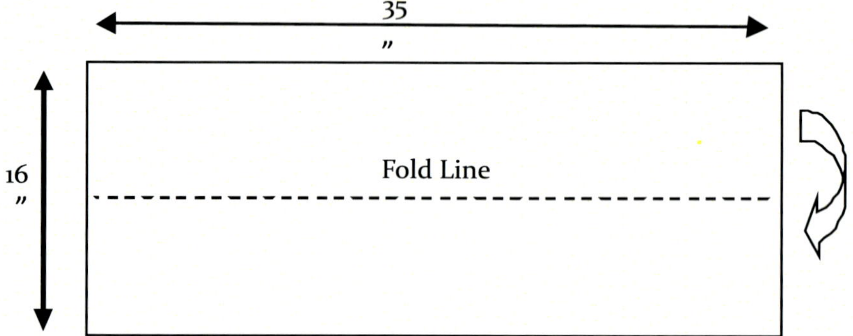

Fig. 2.2

c) Leave a ½" slit at each end so that the casing area remains open and the cord can be inserted into it.

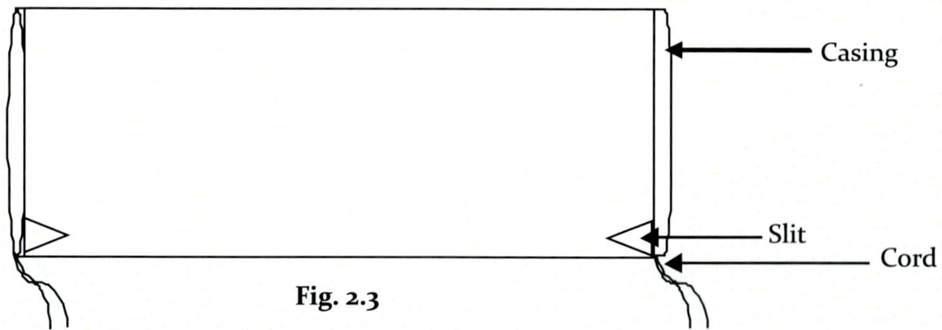

Fig. 2.3

d) This completes the block printed bolster cover made out of an old discarded bed sheet. It is refurbished into a brand new look.

Stuff this cover with a nicely filled bolster and place it interestingly for your comfort on your couch or bed.

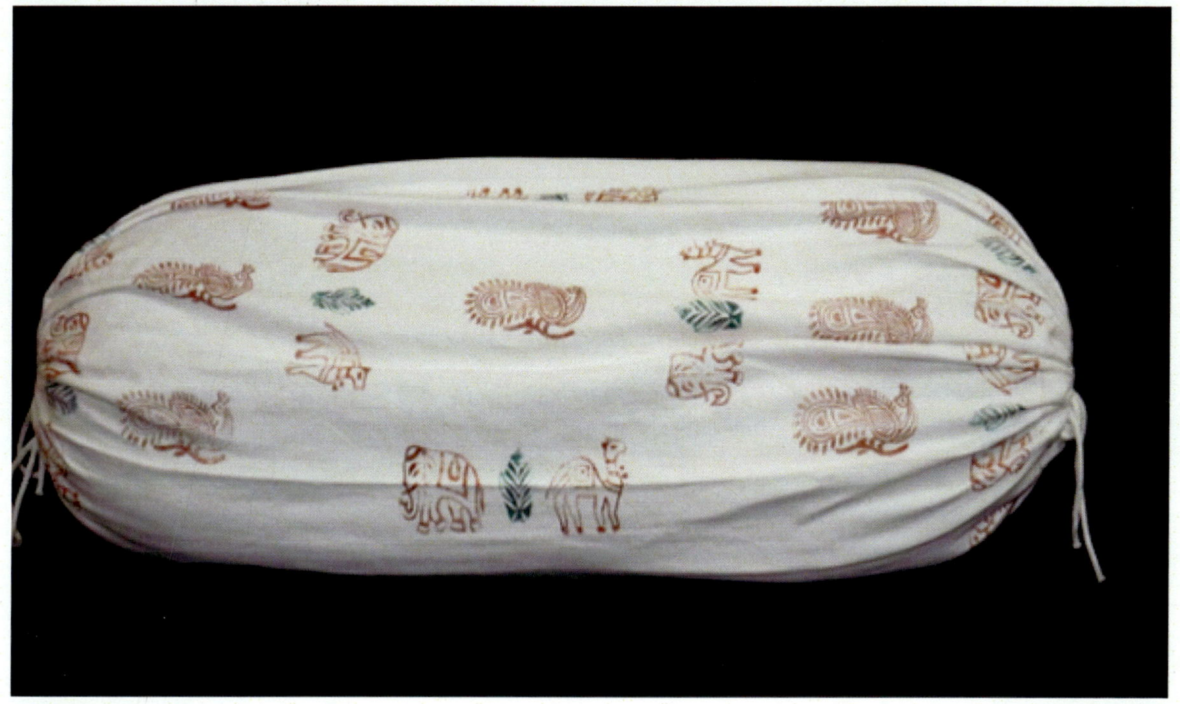

Fig. 2.4 Final Bolster.

My Ingenious Labour Saving Device

I made my own labour saving device for efficiency and accuracy. Block printing is a traditional method of printing fabric to make it look attractive. It can be tedious and cumbersome if the artisan does not have the right equipment and tool to do his job. I have developed a tool that can enable the creative artist to mint the design repeat with accuracy. The cost of this tool is minimal. The technique of block printing in a traditional way of printing requires a lot of effort to place the blocks at regular intervals, without going out of line. My technique yields neatly aligned motifs, printed flawlessly, several in one!! The purpose here is accuracy and speed

Material Required

1) Two wooden blocks of size 2"x2" and one block of size 1.5"x1"
2) 12" long metallic scale
3) 3 screws of 1/2" length and a screw driver

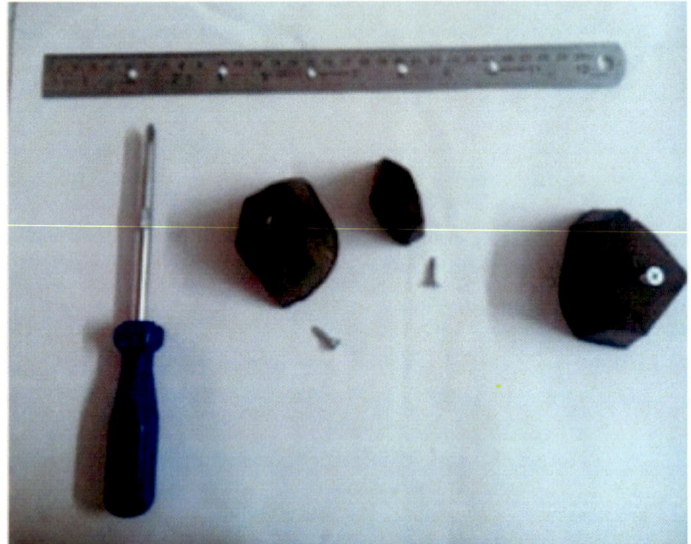

Fig. 1 Wooden Blocks, Screw driver, Screws, Metal Scale

Steps

1) Holes are drilled in metal scale at a distance of 2" apart, starting from any point of an inch.

Fig. 2 Drilling holes in a metal scale.

Fig. 3 Fixing the block to the metallic scale

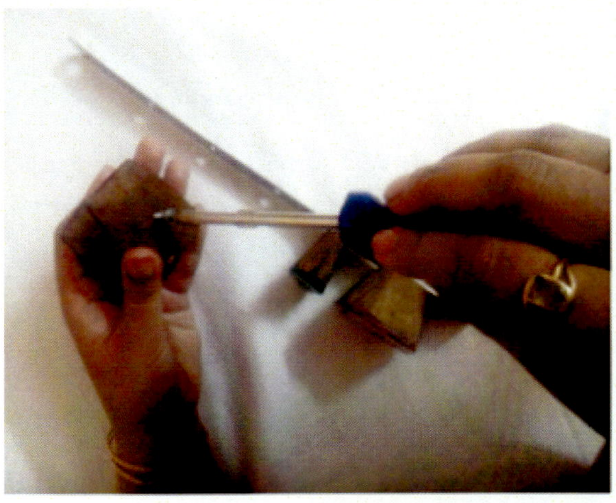

Fig. 4 Fixing the screw to the block

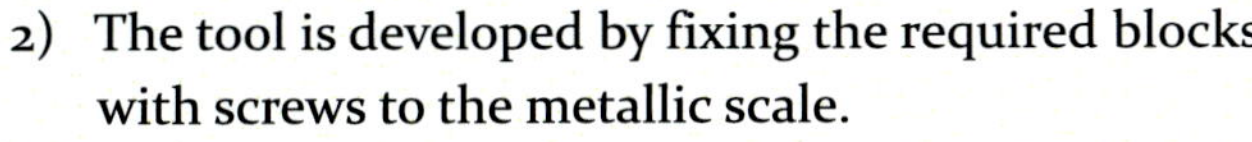

2) The tool is developed by fixing the required blocks with screws to the metallic scale.
3) To nicely align and balance the blocks, pieces of cardboard can be inserted in the space between the blocks and the scale.
4) The metal scale acts as guideline to help the operator work faster and efficiently.
5) The scale also acts as a handle for stamping the block design.
6) This is my ingenious labour saving device. You can make your own!! The tool saves time and increases efficiency allowing the artist to work fast and with precision

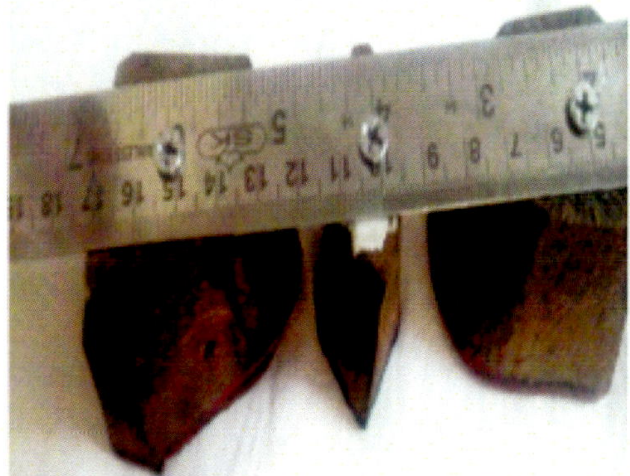

Fig. 5 Inserting a piece of cardboard between the blocks and metal scale to level the device.

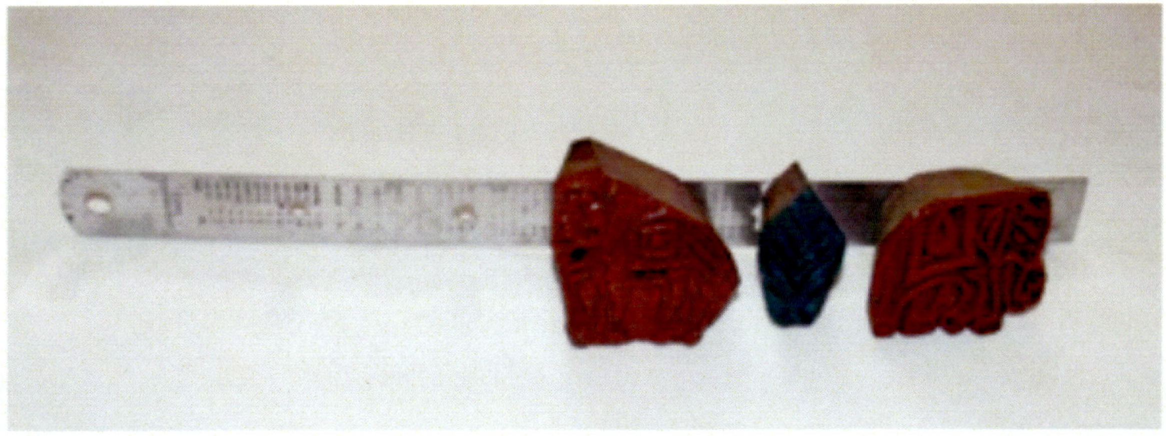

Fig. 6 Ingenious Tool.

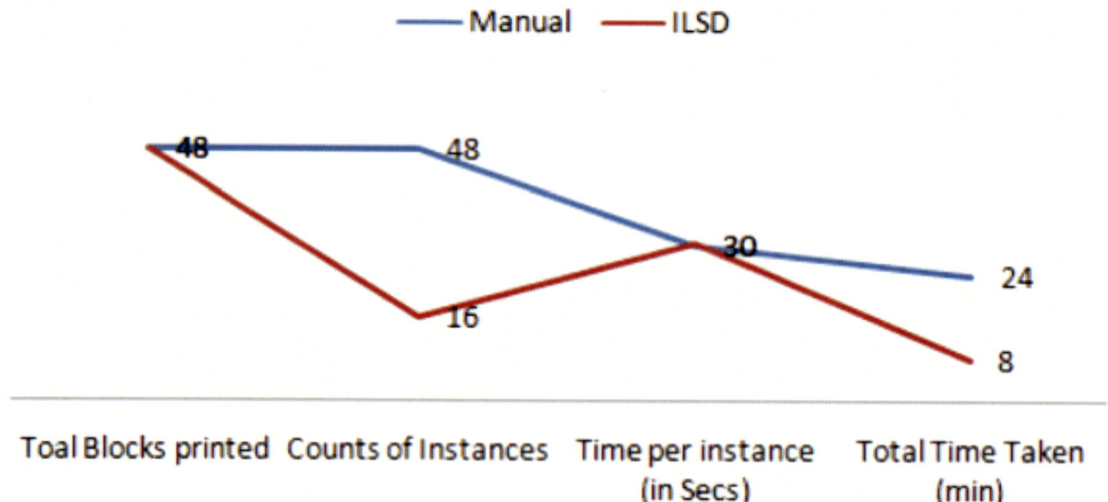

48 | 48 strikes and 16 strikes | 24 minutes; and 8 minutes

Fig. 7 (ILSD Intensive Labour Saving Device)

Fig. 7 shows that if 48 blocks are to be used to print a certain design, it would take 24 minutes to complete the job, printing block by block and only 8 minutes to print with the help of ILSD.

1. It results in accuracy in vertical; horizontal; diagonal alignment of print due to the use of blocks fixed to a scale.
2. There is less wastage of tangible materials like fabric, colour and intangibles like time, effort and labour.
3. It also improves creativity of printing.

Harpreet Kaur
Email: harpreetfd.kmv@gmail.com
Assistant Professor
KMV College, Jalandhar

I love to keep my knick knacks in cardboard boxes of different sizes. It helps me to be more organized with my trinkets. I advice my friends to do the same and avoid having cluttered drawers be it stationery, medicines, make-up items and or any other collectibles.

I do not like to throw out my shoe boxes or any other good size cardboard boxes. I prefer to redo their look in one or the other ways I have mentioned in these pages. A little effort goes a long way in making my storage boxes look new and more interesting.

Products 8, 9, 10 and 11 : Decorative Boxes to Store Trinkets

Product 8. Textile Print Box

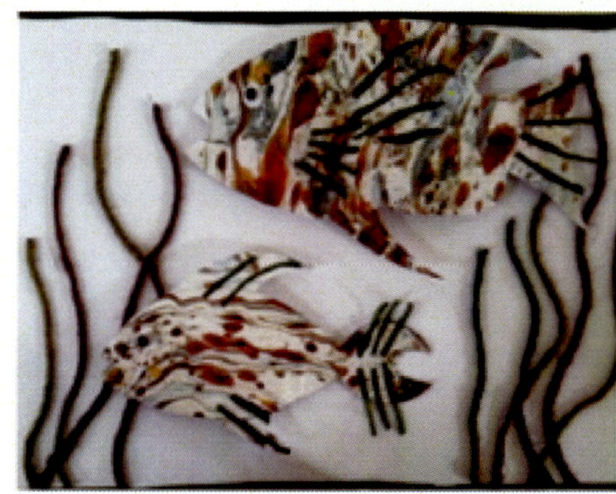

Product 9. Fish Cut Out Box

Product 10. Marble Print Dragonfly Box

Product 11. Dainty Dragonfly Box

Product 8: Textile Print Box

Fig. 1

Material Required:

1. Old discarded piece of printed fabric sourced from your wardrobe
2. Fevicol
3. Scissors
4. Measuring Tape
5. Shoe Box- 4" × 11" × 3.5"

PROCESS

1. Measure the dimensions of the box.

Fig. 2

Fig. 3

2. Neatly iron a piece of discarded fabric.
3. Cut the fabric by calculating as follows:
   - 3.5" (inside depth of box) +3.5" (height of box) + 4"(length of base of the box) + 3.5" (height of box) + 3.5" (depth of box) = 18" is length of fabric
   - 3.5" (depth of box) + 3.5" (height of box) + 11" (width of box) + 3.5" (height) + 3.5" (depth) = 25" is width of fabric

   i.e. , 18"x25" fabric.

   *How much fabric is to be inside the box is entirely your choice. Here I have calculated by taking the complete inside depth and base of the box.
4. Mark the dimensions of the box on the cloth; i.e. the length, width and the base of the box.

Fig. 4

5. Mark points 2" diagonally away from the vertices of the box marking. Draw perpendicular lines i.e., one vertical line beneath it and another horizontal line going away from the box from this point. Cut out the rectangle just formed. Repeat in all the corners.

Fig. 5

Fig. 6. This is how the pattern looks after removing rectangular pieces from all the corners.

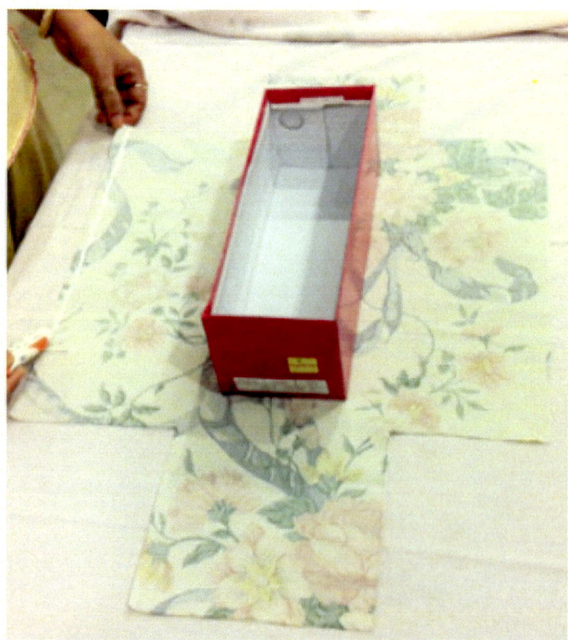

Fig. 7 Choose one of the longer sides of the box and apply fevicol on the edge of the fabric. Apply fevicol to the fabric below the edge of the box too.

Fig. 8 Firmly hold the inside of the box and press down the edge of the fabric along the inner edge of the box. Apply pressure to ensure that the adhesion of the fabric is done firmly and neatly onto the box. Repeat on opposite side.

Fig. 9 Fold the 2" extra fabric along the smaller side of the box neatly.

Fig. 10 Give a diagonal snip of about 2" to the corners of the fabric.

Fig. 11 Glue extra fabric inward to create a seamless edge like shown here. Repeat for all corners.

Fig. 12 Now hold the two parallel edges of the smaller sides and glue the edge of the fabric. Bring the fabric in tightly with one hand while holding the box with the other so that you get a neat edge.

Fig. 12 Glue the edge inside the box. Make sure that the edge of the fabric coincides perfectly with the edge of the fabric. Repeat on the other side.

Fig. 13 The fabric covered Pretty Shoe Box is ready. Repeat the same process for the lid.

Marble Print on Fabric and Ivory Sheet

Material Required

1. Shoe Box of size 8.5" × 13.5" × 8.5"
2. 32.5" × 27.5" Waste fabric to cover the box (1"fold in allowance is included)
3. White Ivory sheet-1 (to cover the lid)
4. Enamel Paints: Red, Blue, Green, Yellow
5. Coloured Tape or Ribbon of 1 cm width for edging
6. Cotton Cord: 55" length; 10mm thickness dyed in red, green and yellow colours
7. Gateway Sheet
8. Double - sided adhesive tape
9. Black beads -2 (5-7 mm for eyes)
10. 4" Blade Paper Scissors
11. 12" Ruler
12. Glue Gun and Glue Sticks
13. 30" Wide Vessel
14. Spoon
15. Stapler
16. Latex Gloves

PROCESS

1. Measure the dimensions of the box and cut the cloth keeping fold-in allowance of 1" on all sides.
2. Staple the ivory sheet and cloth together

Fig. 1

3. Half-fill a 30" wide vessel with water. Sprinkle drops of enamel paint with a spoon on the surface of the water. Add as many colors as desired.

Fig. 2

Fig. 3 Mix the color droplets slowly till you see a marbling pattern of your choice emerging on the surface of water.

Fig. 4 Place one end of the cloth lightly on the surface of water laden with paint and smoothly pull the rest of the cloth along in one soft and quick movement.

Fig. 5 Turn over the stapled cloth-paper pattern and now dip the paper –side in the container .Let the pattern drip dry. (Wear latex gloves if you do not want color on your hands.)

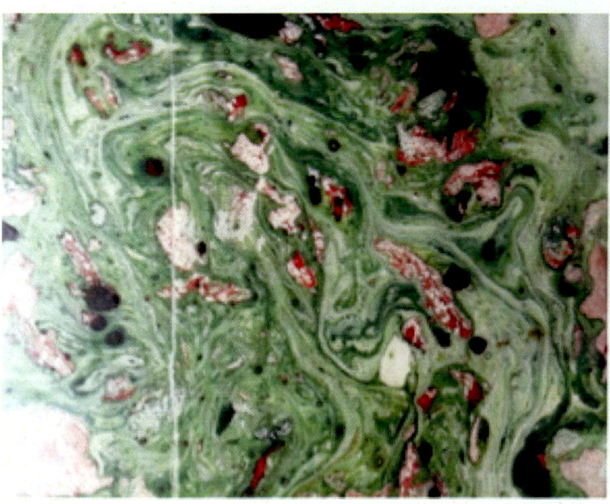

Fig. 6 Beautiful patterns emerge with this technique as it is an ideal way to control colors. Besides, every time there is an element of surprise in the emerging color patterns.

The marble patterned fabric and paper are ready to use as soon as they dry out.

Making Fish Cut Out Templates

PROCESS

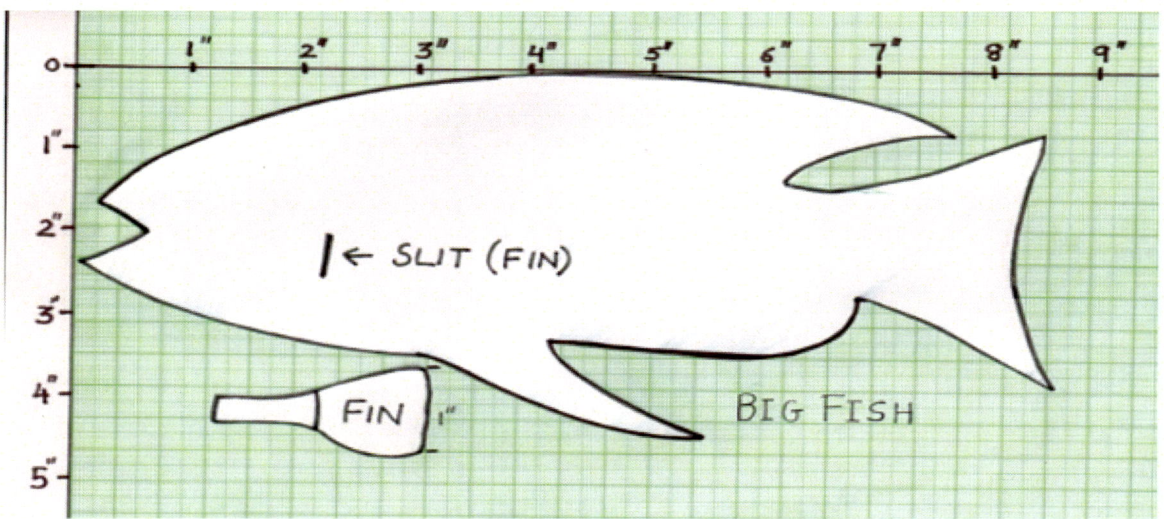

Template 1- Big Fish

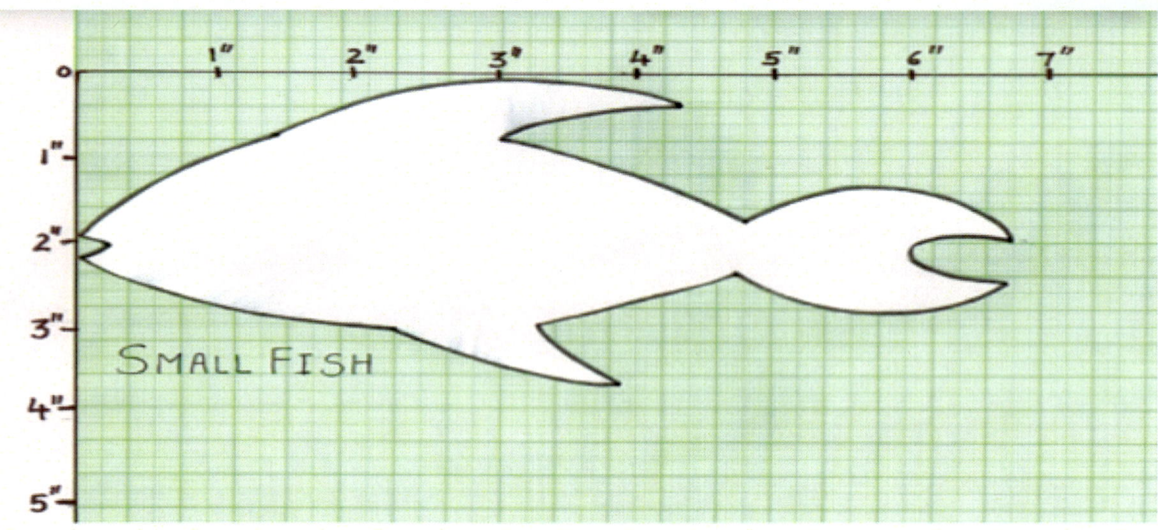

Template 2- Small Fish

Product 9: Fish Cut Out Box

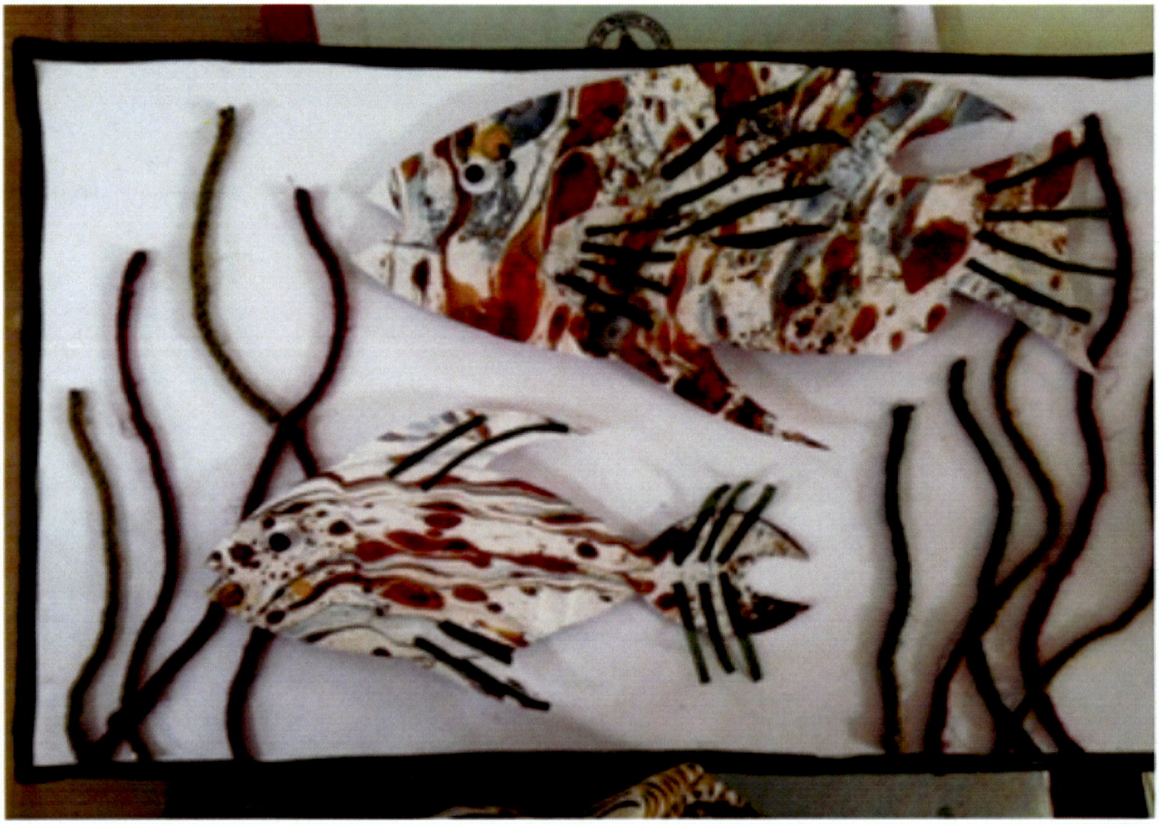

Fig. 1

Highlights of the product-

I have created a marble print on a) fabric and b) paper and used these to cover my box. The fish cut-out shown also has marble print done on fabric. It is pasted on tetron to give it stiffness and a 3D effect.

A box of suitable dimensions is covered with the marble print fabric or paper. Steps to cover the box are the same as mentioned in product 8.This artwork serves as a pretty storage container for collectibles.

PROCESS

1. Cut beautiful multicoloured fish patterns using the fish cut out templates on the textured ivory sheet prepared by the marbelling technique.

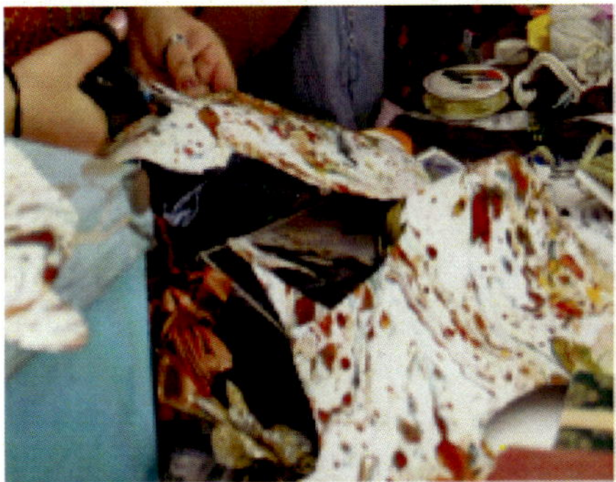

Fig. 2

2. Give a 0.5" slash where the fin is to be attached(as shown in the template). Insert the fold-in section of the fin cut-out inside the slash and fold inside and paste. This gives a three –dimensional effect to the fin.
3. Paste a plain white circle and a round black bead to make an eye.

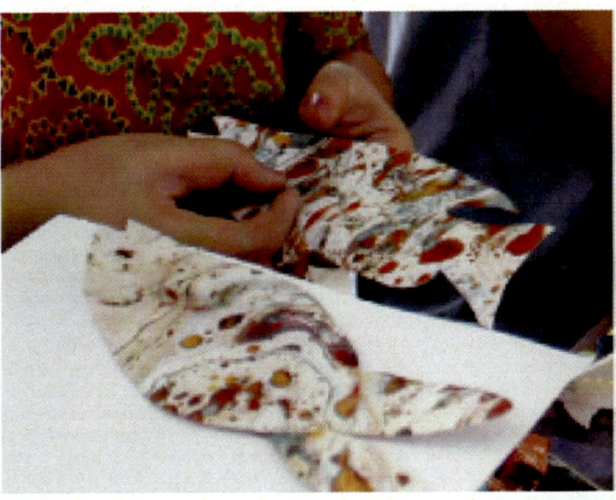

Fig. 3

4. Attach rectangular strips of paper of 0.25 cm of a color of your choice on the fins and tail to contrast with the textured effect.

5. Take a base sheet of size 8.5" × 3.5". Draw weeds on this base sheet with a pencil.

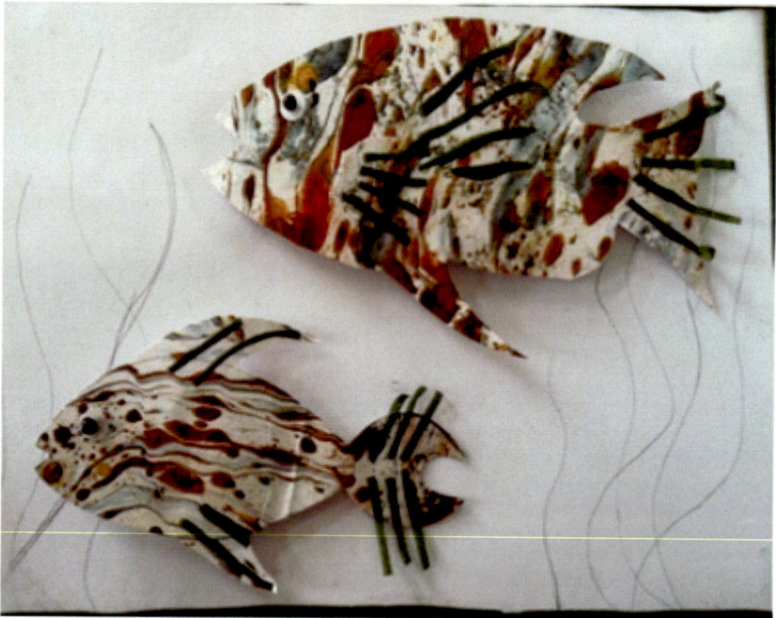

Fig. 4

6. Stick the multi hued cotton cord on the ivory sheet with a glue gun or any suitable adhesive. Take three layers of the double sided adhesive tape and fix them together and attach the fish on them. Fix this bunch of tapes and fish cut out on to the ivory sheet. This gives a 3-D effect to the design. Paste the prepared sheet on the lid of the box.

Fig. 5

Fig. 6

7. Fix strips of textured sheet/cloth of size 2" × 13.5" and 2" × 8.5" on the edges of the lid. To give a finished effect paste a colored tape or ribbon on the edges of the box as well.

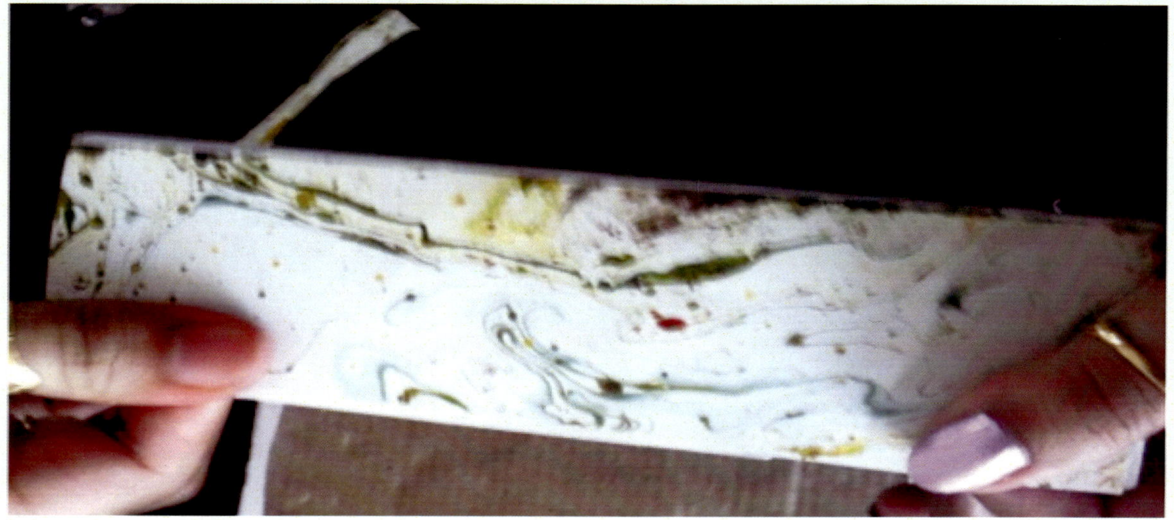

Fig. 7

8. Cut two pieces of marble printed paper or fabric (for the longer side of the box take 13.5"x6" size and for the shorter side take 8.5" × 6" size) and use these to cover the walls of the box.
9. Cut two pieces of marble printed paper or fabric of size 13.5" × 8.5" and paste these at the base of the box (both outside and inside)

The Fish Cut Box is waiting to be used !!

Fig. 8

Products 10 and 11: Twin Boxes: Marble Print Dragonfly Box and Dainty Dragonfly Box.

Fig. 1

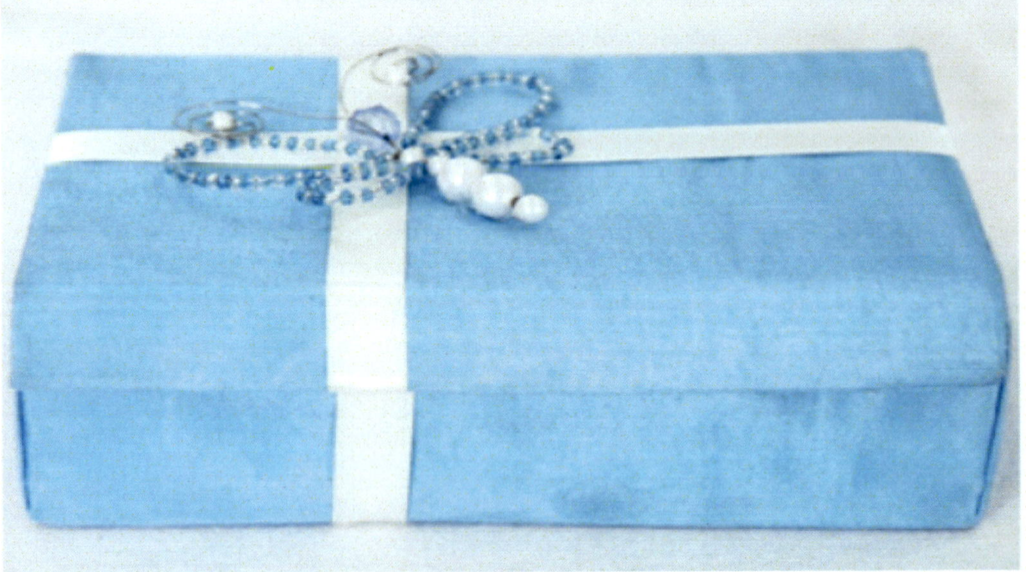

Fig. 2

These decorative boxes are embellished with ribbons and a dragonfly made with an assortment of beads. The first box is covered with textured sheet prepared by marbelling technique (illustrated in Fish Cut Out Box) while the second box is covered with cloth (see process of my Textile Print Box, product 8 above). These dainty boxes are ideally used as jewellery or gift boxes.

Material Required

1. Assortment of beads : 6-7 pearl or plastic beads of 6-8mm
2. Blue, white and purple seed beads 15 gms each
3. Purple bugle and copper seed beads 15 gms each
4. Silver 24 – gauge craft wire: 24" long
5. Ribbons –white (1 cm width) 1 meter; purple (0.5 cm width) 2 meters.
6. 2 Boxes of size 4.5" × 6.5" × 2.5"
7. Textured marble print paper or cloth
8. Blue mercerized cotton cloth 13.5" x 11.5"
9. Wire cutters
10. Ruler
11. Glue gun and glue sticks
12. Paper cutting scissors of 4" blade
13. Fevicol
14. Cloth cutting scissors of 7" blade

Making the Dragonfly

PROCESS
Step 1
Body of Dragonfly

Fold the metallic wire into half; insert a seed bead and twist. String a 6-8 mm bead and twist and then string three more beads of almost similar size to form the body.

Step 2

Lower wings of Dragonfly

String around 30 seed beads onto one wire, then fold it into a loop and give it a twist to form the lower wing. String the other wire similarly to make the second lower wing.

Fig. 2.1

Fig. 2.2

Step 3

Upper wings of Dragonfly

String 40-50 bugle beads on either side of wire to make the upper wings.

Fig. 3.1

Fig. 3.2

Step 4

Head and Antennae Of Dragonfly

Hold the wires together and string an 8mm bead for the head. Twist the wire and cut it with a wire cutter leaving 2" at the end. Place a 6mm bead on either end of the wire and twist the ends of the wire around each bead .Use your fingers to bend each wire into a spiral to form the antennae.

Fig. 4.1

Fig. 4.2

This pretty insect is ready to fly !!

Fig. 4.3

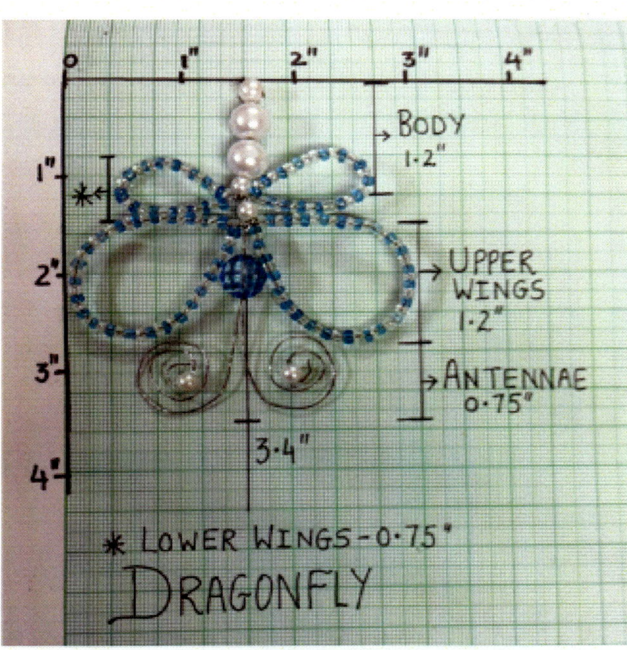

Fig. 4.4 Dimensions of Dragonfly

Product 10: Marble Print Dragonfly Box

Fig. 1

PROCESS

1. Cover the box with marble print ivory sheet using the technique used in Product 2- Fish Cut Out Box.
2. Fix a purple ribbon neatly on the edges of the box with the glue gun. Place the dragonfly leaving 1" from the side and 1" from the top in the left hand corner. Fix into position with glue gun.

Fig. 2

Fig. 3

Product 11: Dainty Dragonfly Box

Fig. 1

PROCESS

1. Cover the box with blue cotton cloth using the technique used in product 8 Textile Print Box shown earlier..
2. Placement of ribbon-
   Mark the position of the ribbon on the box leaving 2" from the side and 2" from the top left hand corner. Fix the ends of the ribbon with the glue gun as shown in the picture.

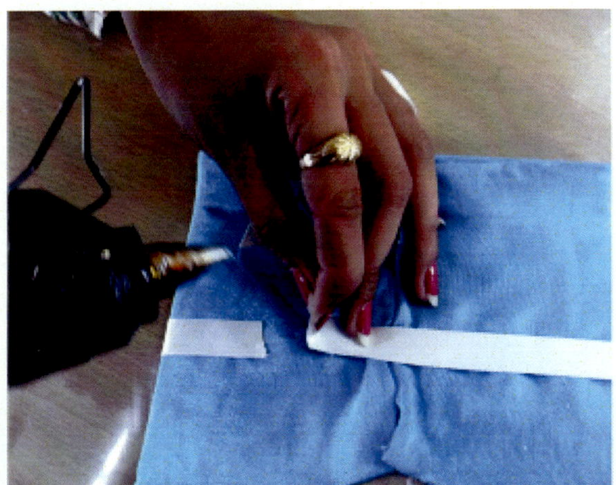

Fig. 2

Fig. 3

3. Attach the beaded dragonfly onto the position where the ribbons intersect with the help of glue gun.

Fig. 4

Fig. 5

The Dainty Dragonfly Box is ready to store your trinkets !

In all these projects I have used discarded shoe boxes and refurbished them with my ideas of artwork. The process was engrossing and stimulating for me. Such projects are an endeavour to make us sensitive to the environment and reduce our carbon footprint. I found it to be a purposeful activity which required patience and consistency in following the steps accurately with determination and sensitivity. I recommend my friends to try out this project and use the boxes thus created for storage and aesthetic appeal

1. References

- Benson, Ann, The New Beadweaving, Sterling Publishing Co., Inc. New York, NY. 2003
- Big Book of Beautiful Beads, by Krause Publications. 2003
- Crabtree, Caroline & Stallebrass, Pam, Beadwork A World Guide, Thames and Hudson Ltd, London, 2002
- Dickinson, Gill. Gift Wrapping For Every Occasion, Quintet Publishing Limited, London, 1994
- Designer Crafts, Anness Publishing Ltd, London, 1997
- Hacker, Katie, Simple & Stylish Bead Accents, Published by KP Books, 2005
- Nehring, Nancy, Embellishing with Beads, Sterling Publishing Co., Inc. New York, NY, 2003
- Quantum Books, Beads Buttons & Bows, Publishing Chart well Books New Jersey, USA. 1997

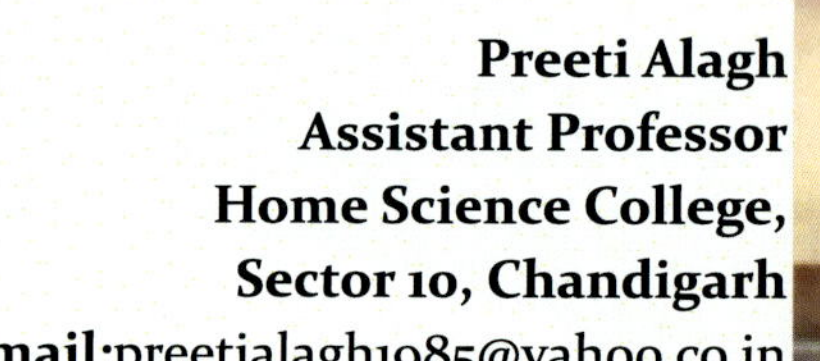

Preeti Alagh
Assistant Professor
Home Science College,
Sector 10, Chandigarh
Email:preetialagh1985@yahoo.co.in

I made this fashionable clutch purse to hold my driving licence and my car keys and not to forget my credit card which I keep losing!! I used the left over ribbons from an Apparel Industry as I wanted to focus on sustainable and green lifestyle products while being creative and ingenuous.

Product 12: Clutch Purse

Fig. 1 Front of Clutch Purse

Fig. 2 Back of Clutch Purse

Material Required

1. Two matching golden color brocade and organza ribbons of half inch width: 3 metres each to make woven pattern (Industrial waste)
2. Velvet Fabric for outer layer of 12" × 14" (Industrial waste)
3. Fuse paper of 25 cms to paste woven pattern
4. Taffeta fabric for inner layer
5. 12" × 12" Tetron to give stiffness
6. Polyster sewing thread
7. Tafetta material of 8" × 10" for pocket

PROCESS:

1. Making up an attractive woven ribbon surface
2. Making up the body
3. Making up the flap
4. Making the gusset and attaching it
5. Attaching the flap to the body
6. Making a pocket
7. Making a bow
8. Finishing up

Step 1 Making Woven Ribbon Surface

My clutch purse has a pretty "woven with ribbons" flap. The rest of the purse is finished with velvet fabric.

Fig. 1.1

Fig. 1.2

For this flap, I took a 5" × 10" size fuse paper and brocade and organza ribbons; these both measured 3 metres each.

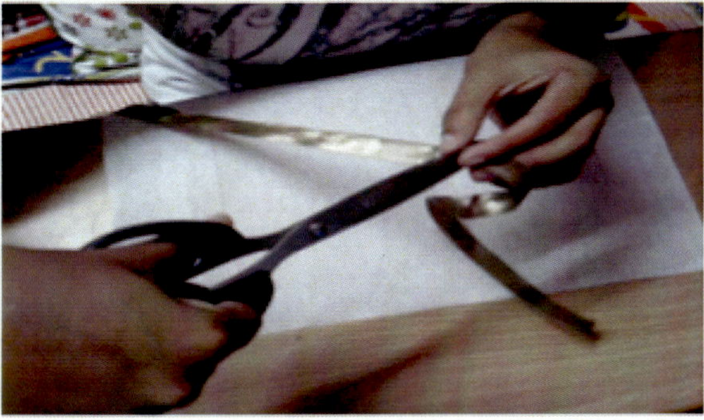

Fig. 1.3

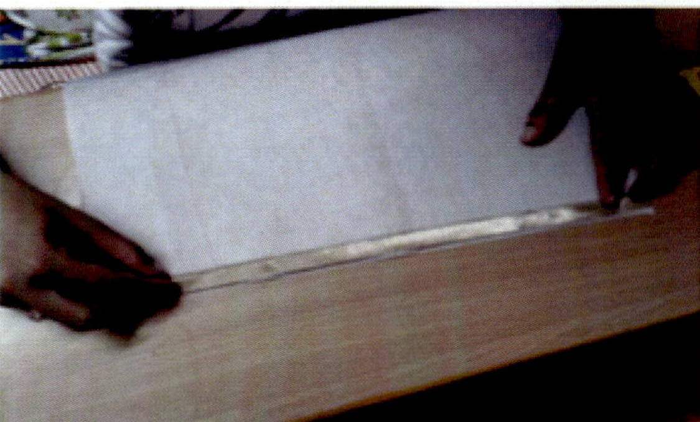

Fig. 1.4

I cut the organza ribbon into 5" long pieces to use for warp woven pattern. Nine pieces were cut.

Fig. 1.5

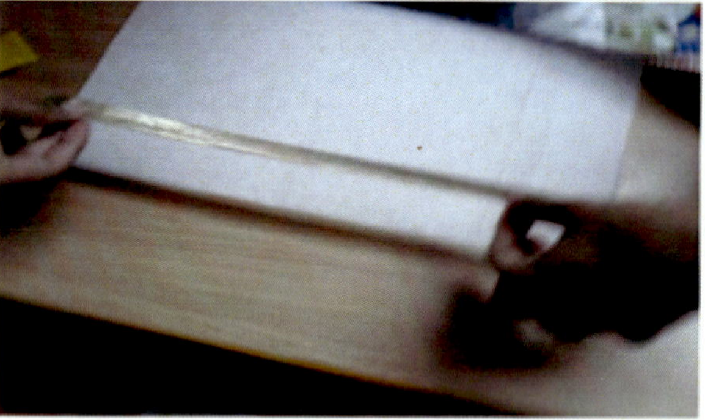

Fig. 1.6

I applied a drop of an adhesive on the corners of the ribbons and pasted them on the fuse paper.

Fig. 1.7

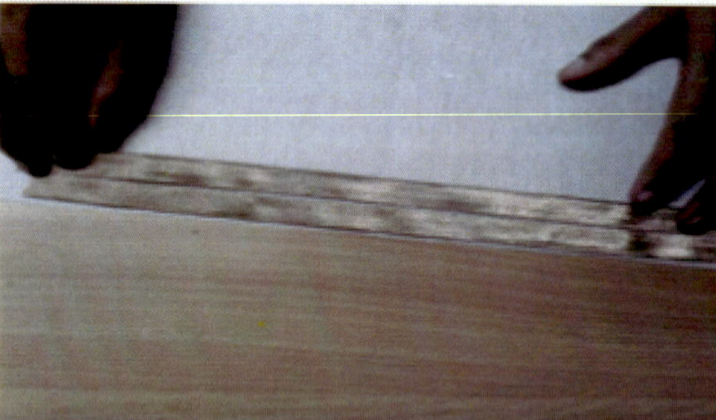

Fig. 1.8

I repeated the process till all the warps got pasted on the fuse paper.

Fig. 1.9

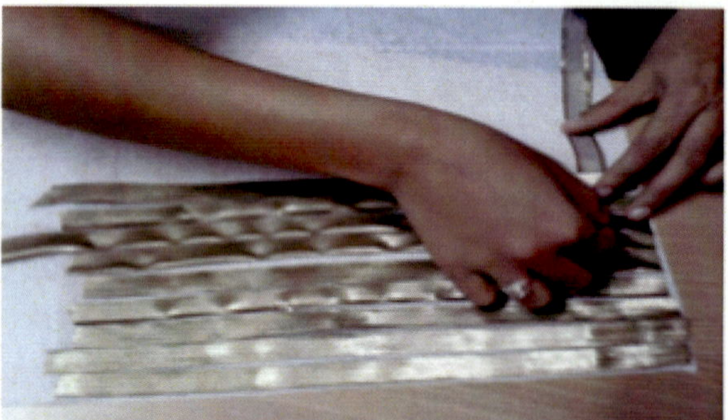

Fig. 1.10

Thereafter, I started inserting brocade ribbons for weft by lifting the organza ribbons up and down.

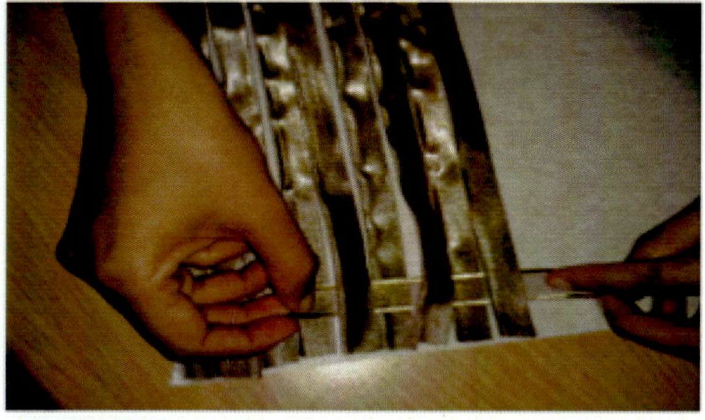

Fig. 1.11

Fig. 1.12

Weft was inserted as shown above.

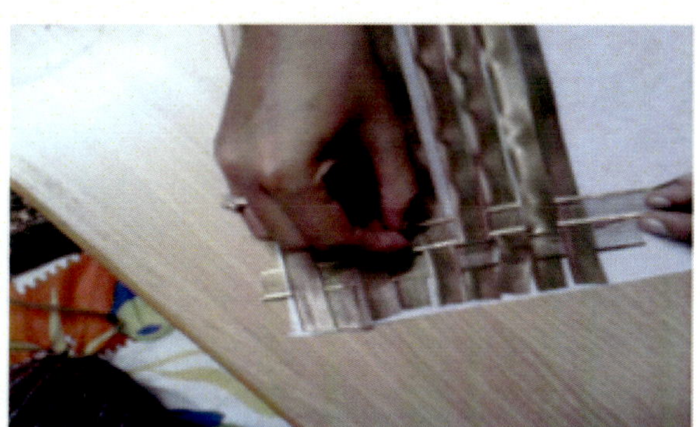

Fig. 1.13

Fig. 1.14

Insertion of weft continued till complete pattern was developed.

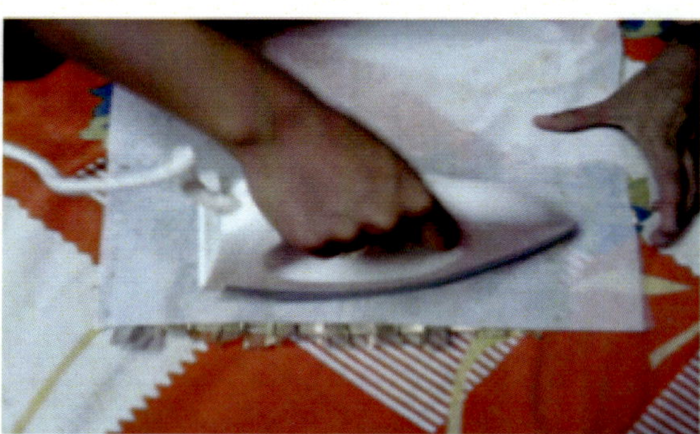

Fig. 1.15

Fig. 1.16

Once the woven pattern was ready for the flap of the clutch purse I ironed it from the back side so that ribbon surface created got pasted onto the fuse paper.

Step 2 Making up the Body

The body has velvet outside, taffeta inside and tetron in the middle layers for stiffness. For this, a template was cut on paper.

(scale: 1/4th)

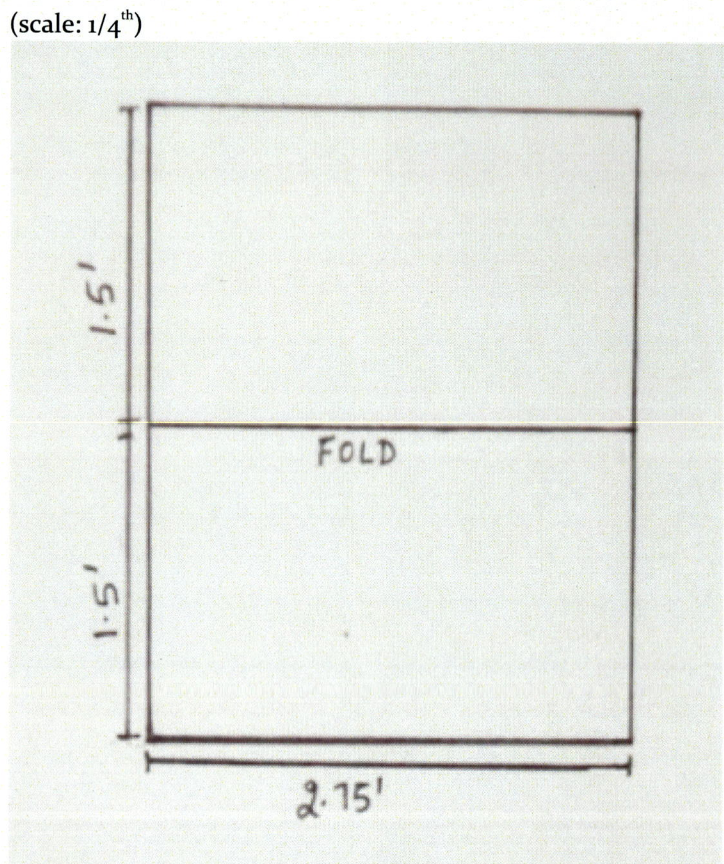

Fig. 2.1

Fig. 2.2

Fig. 2.3

I measured the front and back of an old discarded clutch purse and marked these dimensions 12"×11" on a fuse paper.

(scale: 1/4th)

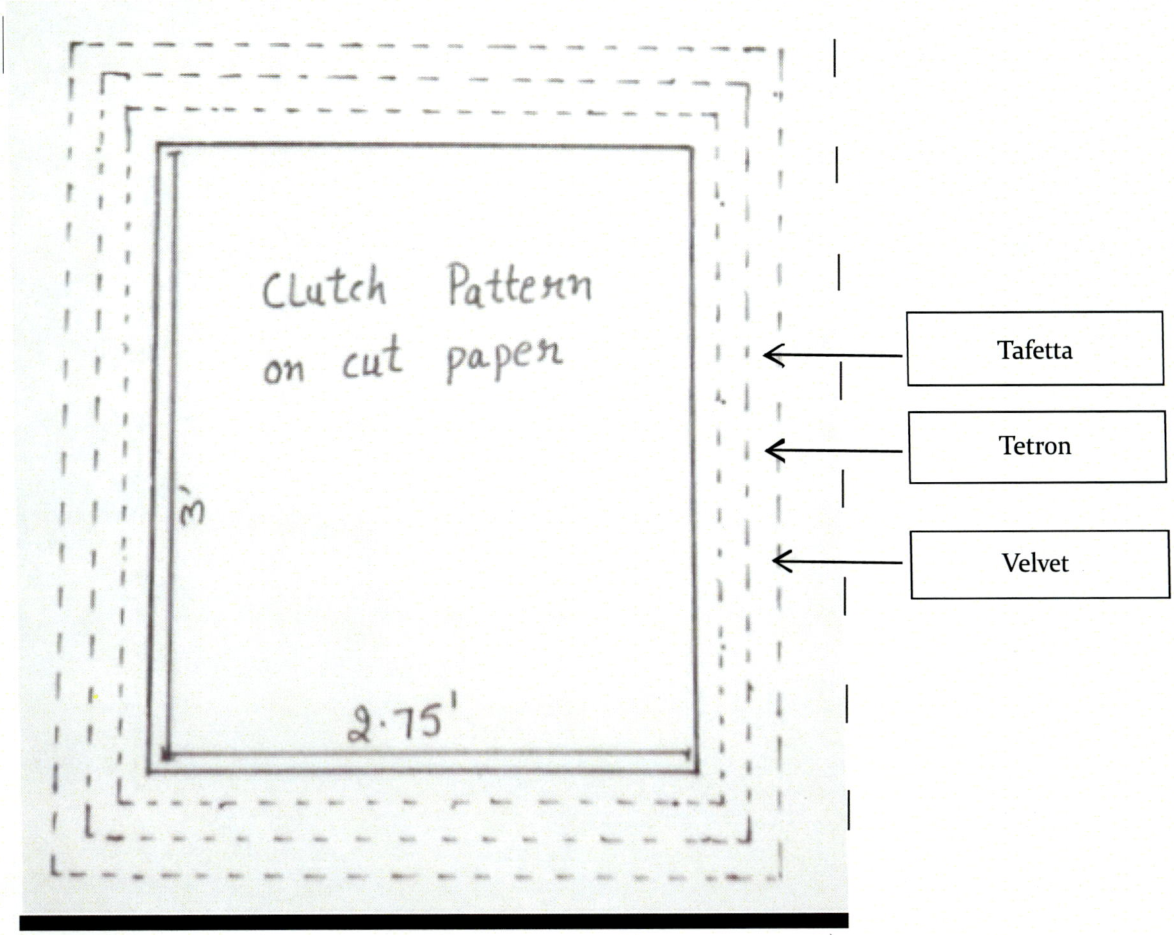

Clutch Pattern on cut paper

3'

2.75'

Tafetta

Tetron

Velvet

Fig. 2.4

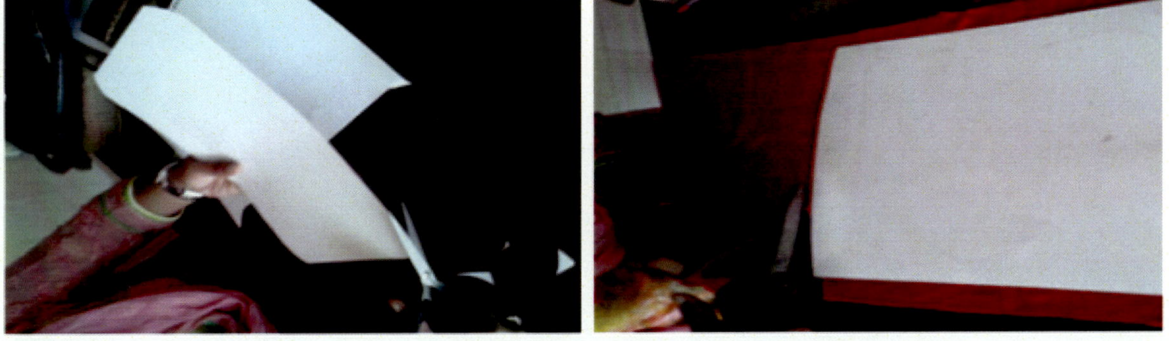

Fig. 2.5

Fig. 2.6

I placed the pattern-cut on three layers as show in figure: Tetron, Taffeta and Velvet Fabric.

Fig. 2.7 These layers were ironed together to flatten out any bulges and wrinkles.

Step 3
Making up the Flap

(scale: $1/4^{th}$)

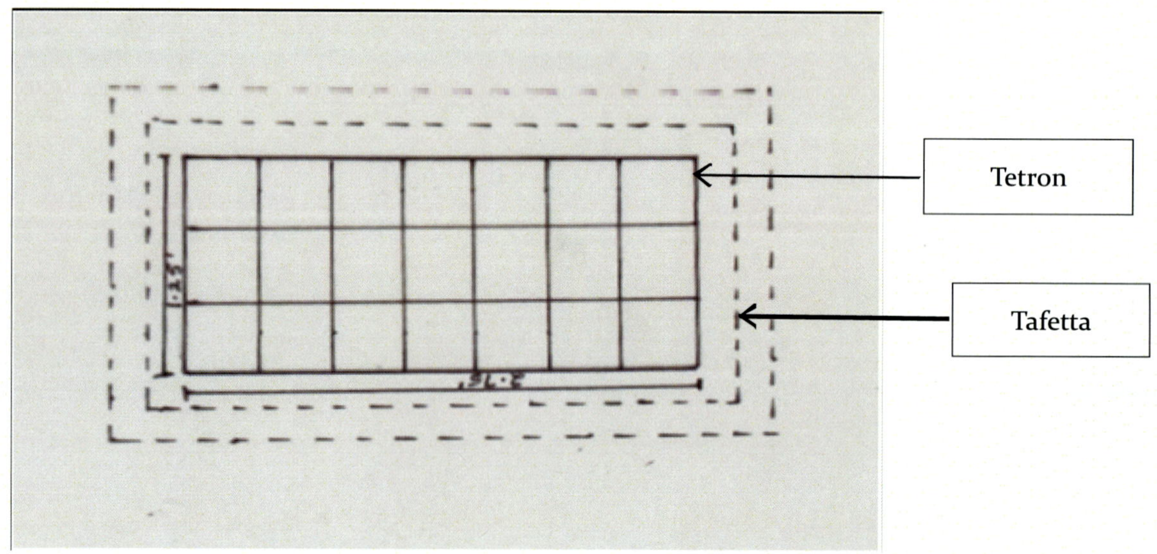

Fig. 3.1

I, cut tetron and taffeta in the rectangular shape for the flap and placed the woven ribbon pattern on taffeta side.

Fig. 3.2

I placed the "woven with ribbons" flap on these two layers: tetron and taffeta and tacked them together.

Step 4

Making up the Gusset

I made the gusset draft on the paper

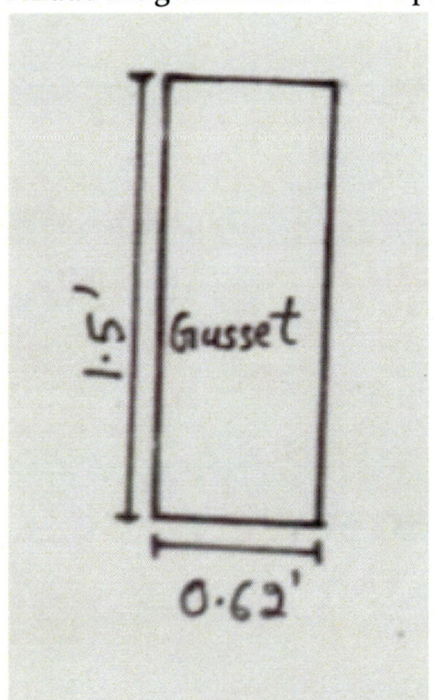

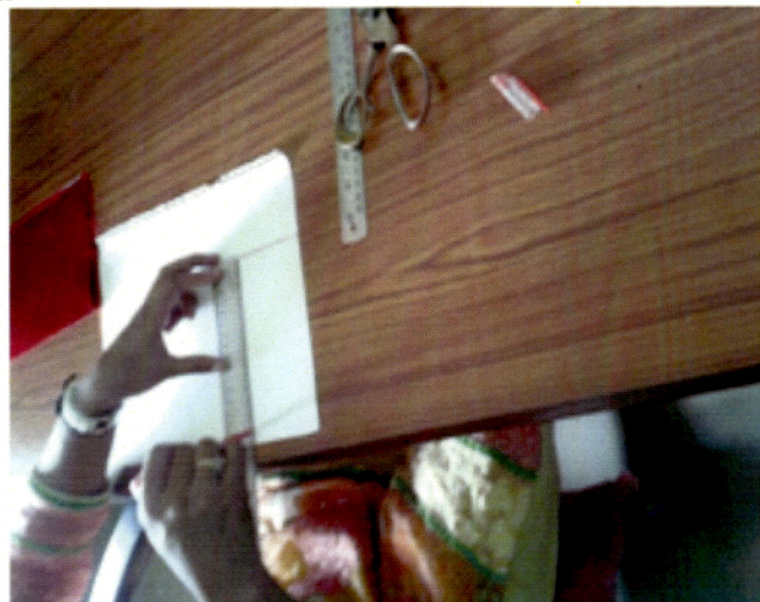

Fig. 4.1

Fig. 4.2

A gusset is a triangular or square piece of fabric inserted between two seams to add breadth or reduce stress.

(scale: 1/4th)

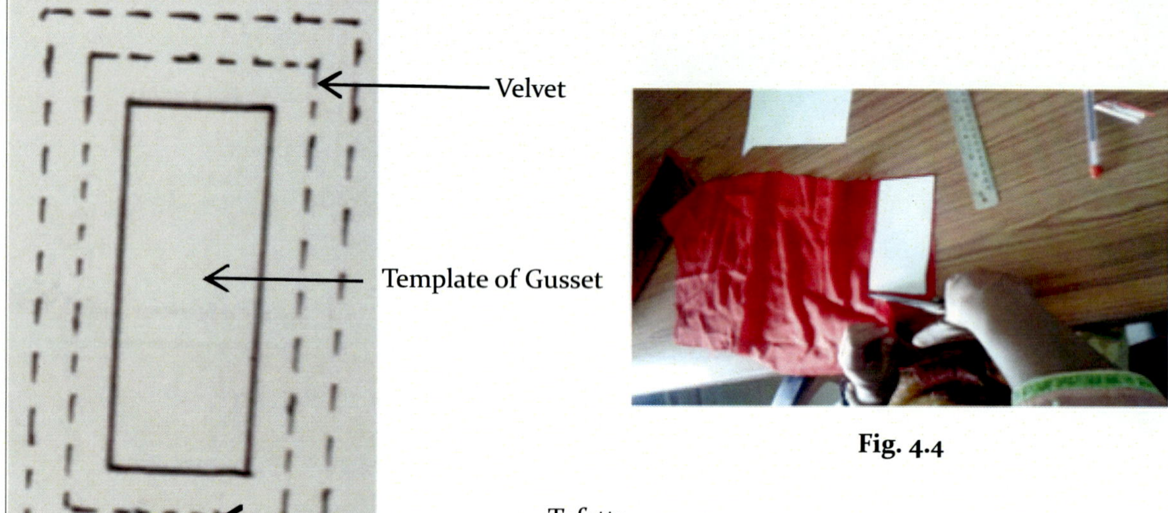

Fig. 4.3

Fig. 4.4

I placed gusset draft on the two layers i.e. taffeta and velvet fabrics.

Fig. 4.5

Fig. 4.6

This was followed by cutting gusset pieces in velvet and taffeta and stitching the gusset to front and back of the clutch purse.

(scale: 1/4th)

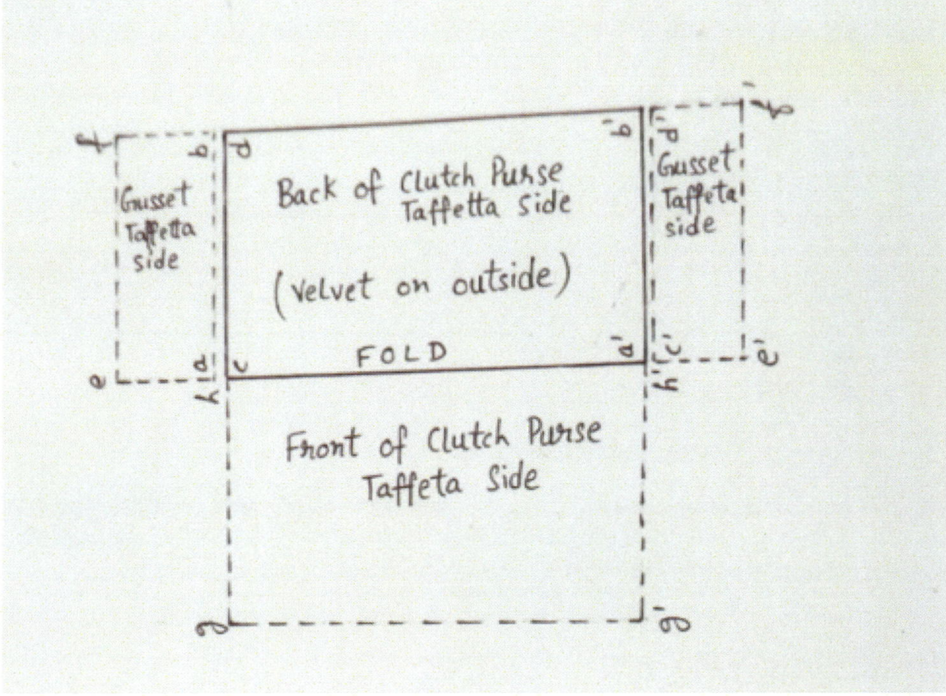

Fig. 4.7

First of all, stitch the three layers of front and back i.e. Taffeta, Tetron and Velvet together with 4 spi (stitches per inch).

Stitch the gusset to the body by joining ab with cd and a'b' with c'd' for gusset. Now, attach other side of the gusset with front as ef with gh and e'f' with g'h'with 6 spi.

Fig. 4.8

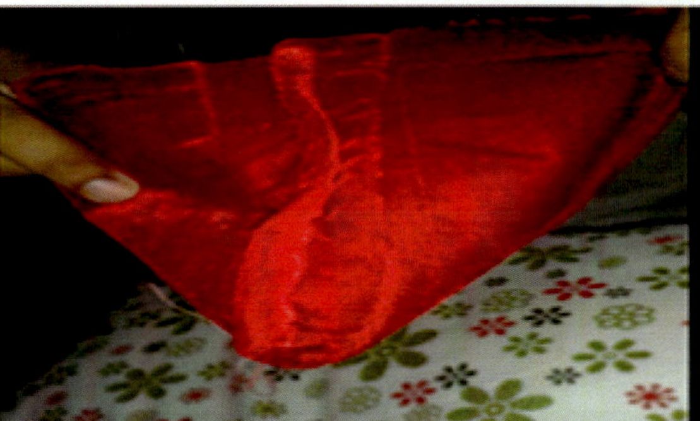

Fig. 4.9

Step 5

Attaching the Flap to the Body

I prepared the flap piece by stitching it on 3 sides and letting one side remain unstitched. This was the side that was to be attached to the back of the clutch purse.

(scale: 1/4th)

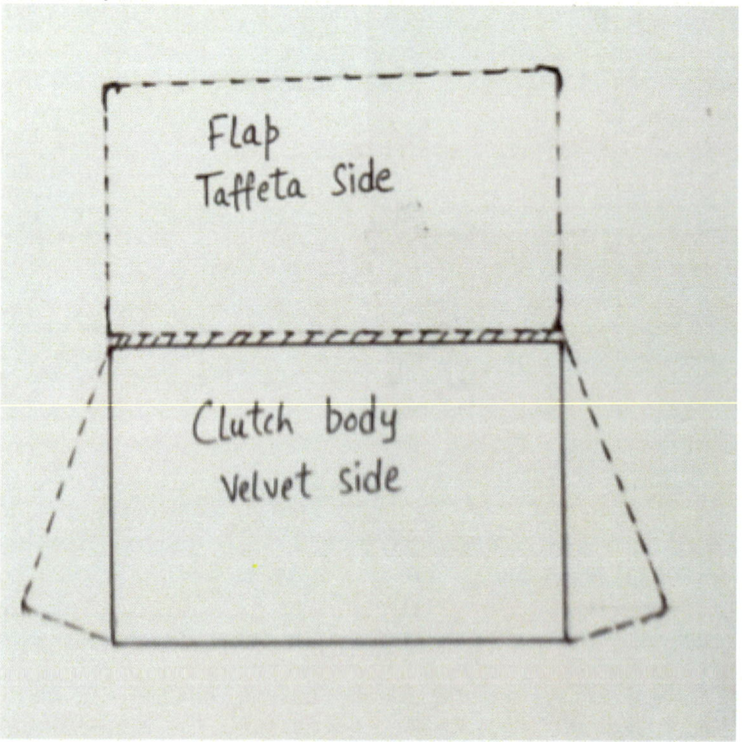

Fig. 5.1

Holding the flap with right side (i.e. velvet surface of the body and ribbon surface of the flap) surfaces face to face. I stitched and attached the flap to the body of the clutch. I did not stitch the taffeta layer but finished this layer with neat invisible hemming.

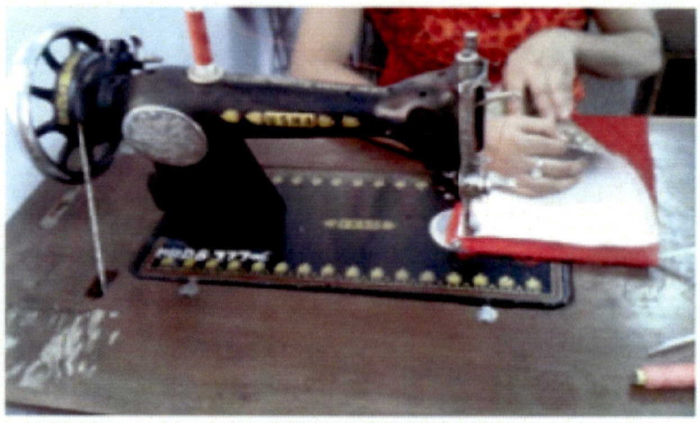

Fig. 5.2

(scale: 1/4th)

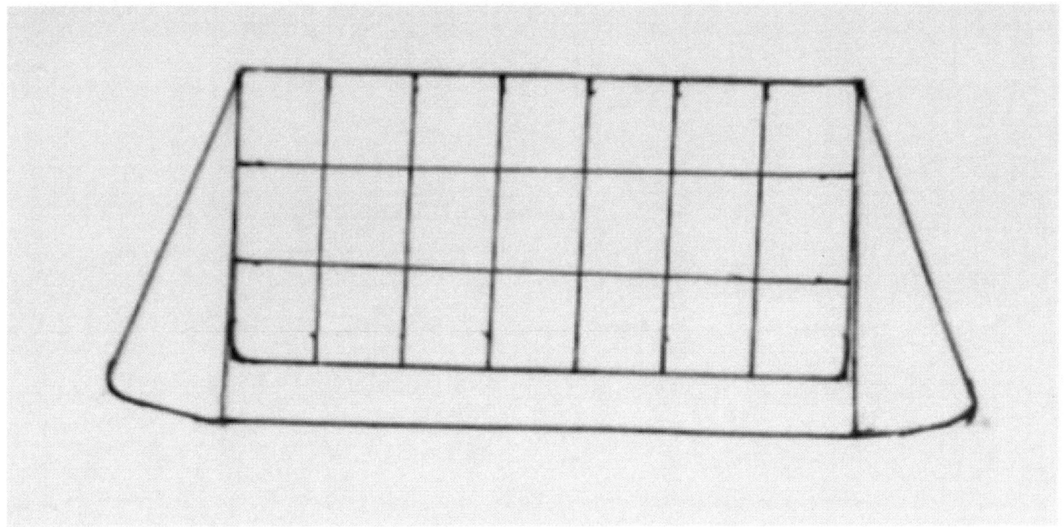

Fig. 5.3

Step 6

Making a Pocket

This is how, I made the pocket.

(scale: 1/4th)

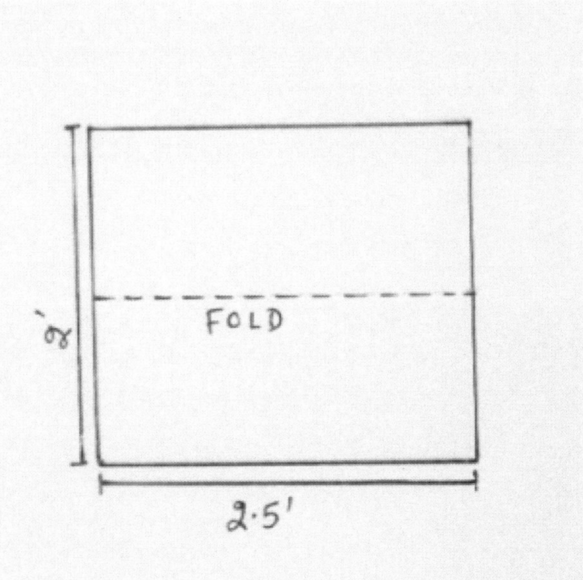

Fig. 6.1

Cut the 8" × 10" taffeta material for the pocket.

Attach the zipper as an "in-seam zipper "using a zipper foot. (Refer to the diagram on next page)

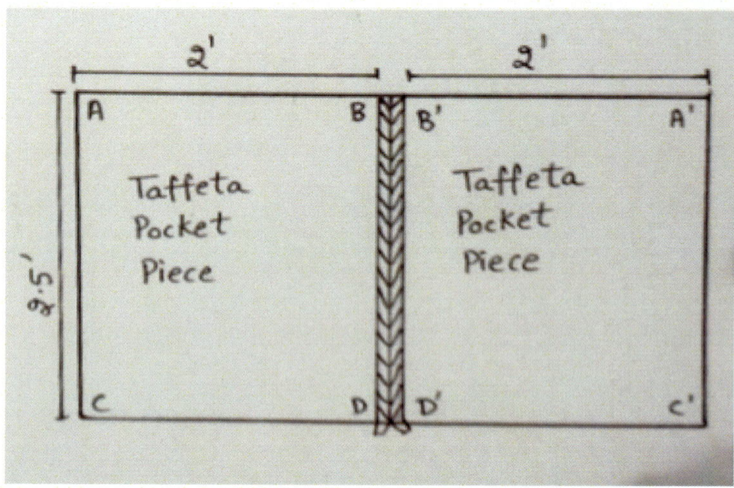

Fig. 6.2 After attaching the zipper, stitch the two sides AB, CD to A'B' and C'D' respectively.

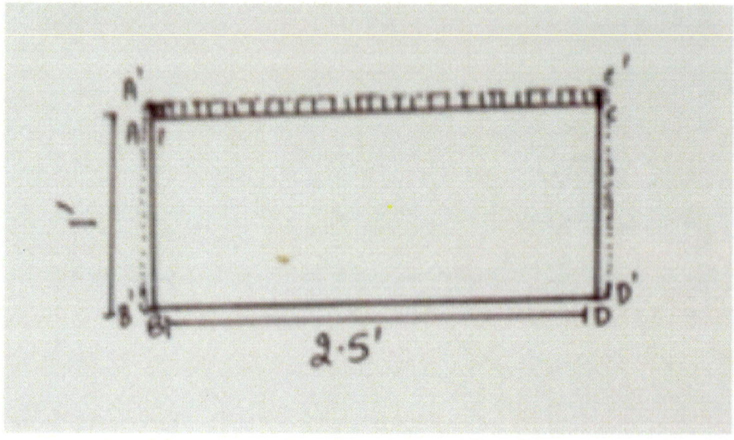

Fig. 6.3

Attach the pocket inside the Clutch Purse by stitching it to the inside with invisible stitches with a hand sewing needle.

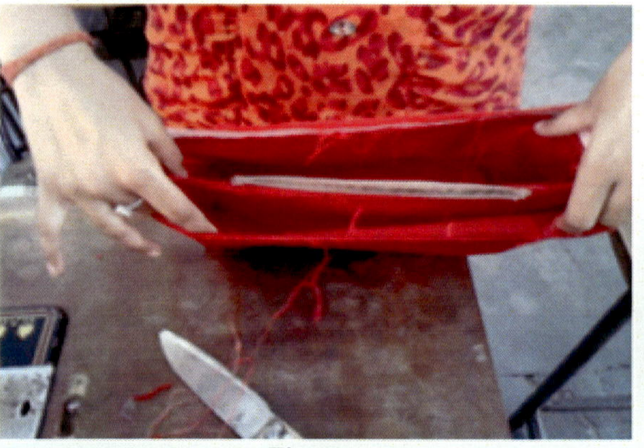

Fig. 6.4

Fig. 6.5

Step 7

Making a Bow

Cut velvet fabric of size 12" × 6".Finish its sides by folding over 1/4" and then folding 1/4" again. Move the machine carefully so that the fabric sits flat and neat. Fold the 12" side of the fabric into half and iron it.

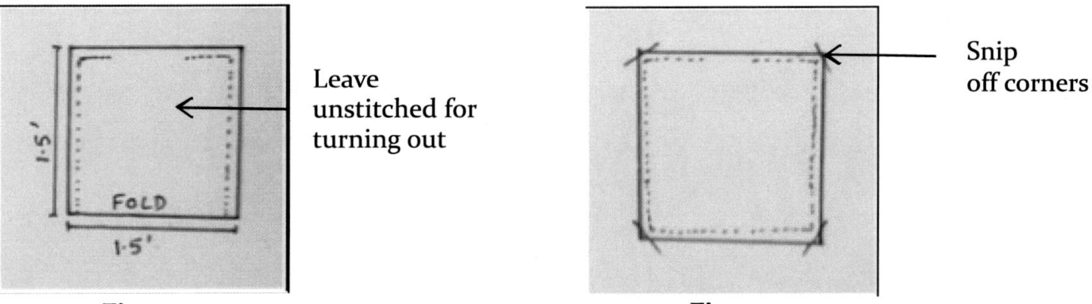

Fig. 7.1

Fig. 7.2

Sew along the side as shown in the picture above. Remember to reverse the stitches near the opening. Slash the seam allowances little bit so that it won't be bulky. Cut the corners off at a 45 degree angle. Then, turn it around and iron it. Top stitch it

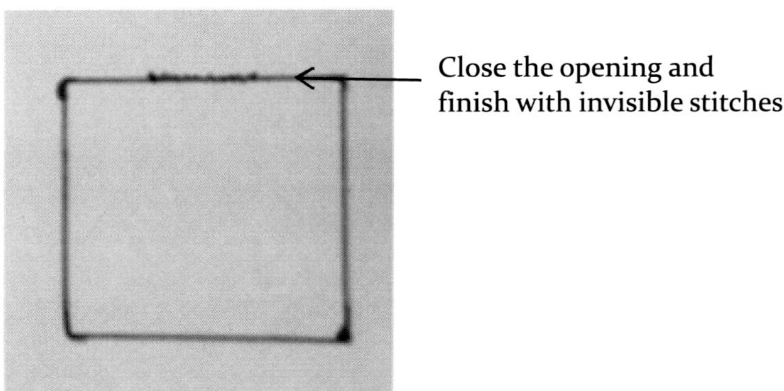

Fig. 7.3

Piece turned right side after snipping off corners.

Iron and crease a centre line of this piece. Run a machine stitch along the pressed line and knot the thread end. Pull the sewing thread and gather up to give the shape of a bow for the clutch purse.

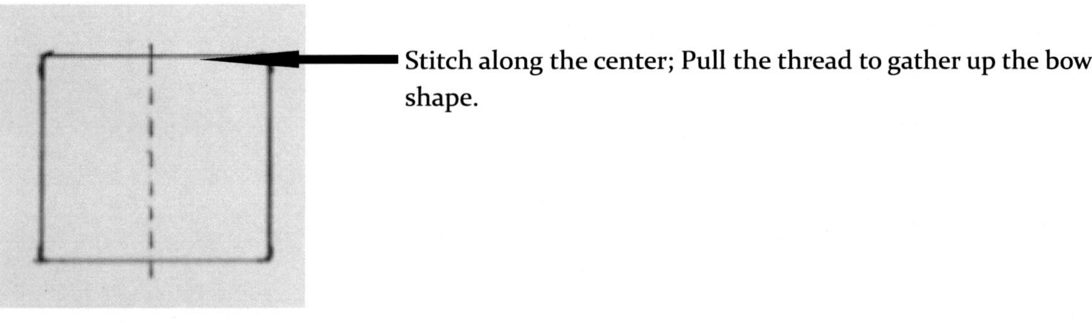

Fig. 7.4

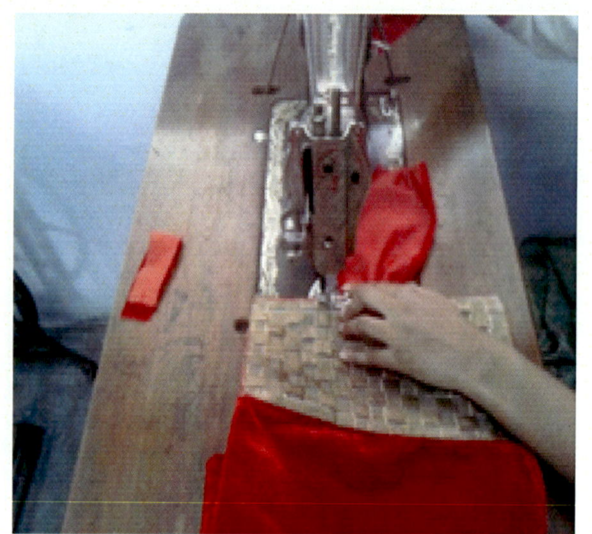

Fig. 7.5

Fig. 7.6

Fig. 7.7

Fig. 7.8

Step 8 Finishing Up

Attaching the binding

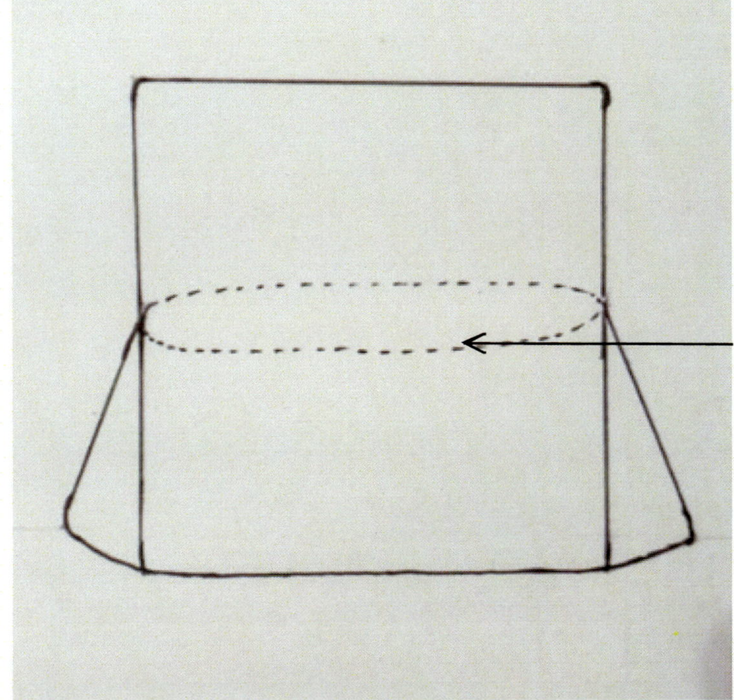

Attach binding around this opening to finish up the raw edges.

Fig. 8.1

The top opening of the Clutch is finished with binding. Cut a bias strip of Taffeta fabric 12" in length and 1" in width to finish the centre edge of the purse as shown in the figure. Keeping the right sides of binding and the stitched clutch facing each other, stitch along the sides. Now, turn the binding inside and stitch along the whole area.

Fig. 8.2

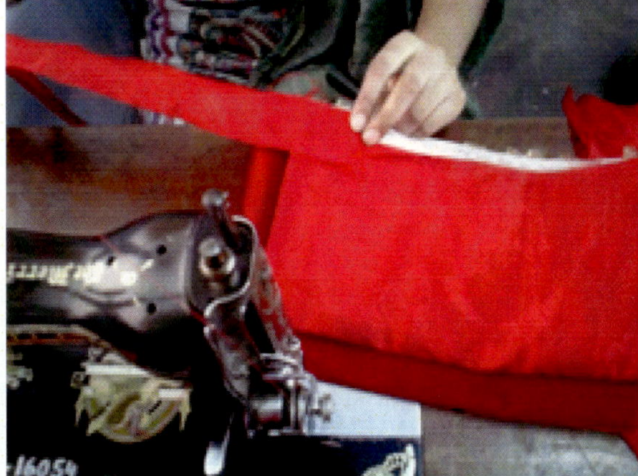

Fig. 8.3

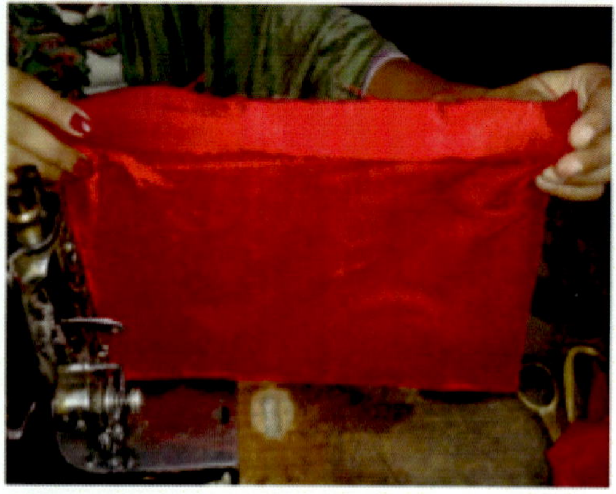

Fig. 8.4

Fig. 8.5

Fold the binding inside and finish it properly.

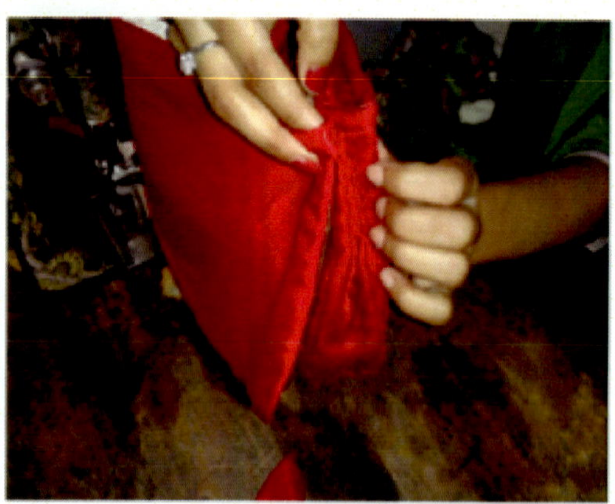

Fig. 8.6

Fig. 8.7

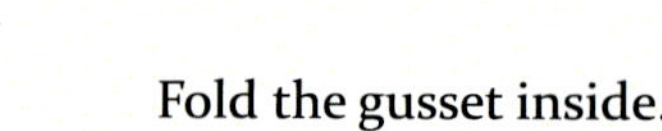

Fold the gusset inside.

Fig. 8.8

Fig. 8.9

Finally, I neaten up by cutting all hanging threads and lint etc.

Fig. 8.10

My Clutch Purse is almost ready; I just need to fix a decorative button in the centre of the bow to make it look more beautiful.

By the way, I took this decorative button from one of my discarded dresses.

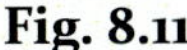

Fig. 8.11

Fig. 8.12

Shweta Sharma
Email: arora_shweta@rediffmail.com
Assistant Professor
N.I.I.F.T. Mohali Punjab

A Personal Lamp to light up my space.

Here is a project designed to sensitize creative minds towards a "Green Environment". The plastic that continues to poison the earth is indestructible but can be molded into useful products, instead of crowding up the landfills and choking them.

When Plastics are mentioned, we tend to think of Plastic Bags, Plastic Bottles and Plastic Sheets! Few of us are aware that Plastic is NON-BIODEGRADABLE, and will never go away from the face of the earth. We are enslaved to this evil product that is a health hazard.

It is important that we recycle plastic, instead of continuously producing more and dumping it in the environment. Plastic is made from Polyethylene Terephalate, PET in Short. Used Pet Bottles are being utilized to produce polyester fiber and technical textiles by Reliance Industries Ltd. Here I have tried to convent PET material into a pretty looking lamp.

Plastics can be Recycled?

A lot of plastic material can be recycled. Most plastic items are labeled on their base with a stamp that shows a triangle made of an arrow, with a number in the middle. The arrow indicates that they can be recycled and the number indicates the type/grade of plastic.

Plastic Grades and Usage

1 PETE	**Polyethylene Terephalate Ethylene** PETE goes into soft drink, juice, water, detergent, and cleaner bottles. Aslo used for cooking and peanut butter jars.	5 PP	**Polypropylene** PP goes into caps, disks, syrup bottles, yogurt tubs, straws, and film packaging.
2 HDPE	**High Density Polyethylene** High Density Polyethylene HDPE goes into milk and water jugs, beach bottles, detergent bottles, shampoo bottles, plastic bags and grocery sacks, motor oil bottles, household clearners and butter tubs.	6 PS	**Polystyrene** PS goes into meat trays, egg cartons, plates, cutlery, carry-out containers, and clear trays.
3 PVC	**Polyvinyl Cloride** PVC goes into window cleaner, cooking oils, and detergent bottles. Aslo used for peanut butter jars and water jugs.	7 OTHER	**Other** Includes resins not mentioned above or combinations of plastics.
4 LDPE	**Low Density Polyethylene** LDPE goes into plastic bags and grocery sacks, dry cleaning bags, flexible film packaging and some bottles.		Technical Textiles are also made from plastic materials.

Fig. 1

As responsible humans let us not consume the earth and rip it off its resources!!!! Let us abide by the 3R's Principle

Fig. 2

Keeping the above in view, I took up this project of creating A Personal Lamp

This is an open ended art idea that can be taken further by the interested community, to make their own accessories, pen holders, planters etc.

Fig. 3 A Personal Lamp

Materials required

1. 5 Used Soft drink bottles/plastic bottles
2. Cutter (blade with a handle)
3. Broad base candle
4. Matchbox/lighter
5. Measuring tape
6. Fevikwick-2 tubes, 5 gm each
7. Bulb and holder with a base-1
8. Tweezers-1
9. Permanent marker

Fig-4

PROCESS

Step-1
Collect Bottles

a) Before starting out with the project collect some empty plastic bottles of different sizes/colors
b) We need to keep in mind the color scheme of the bottles for an aesthetic appeal of the lamp.
   In this project I have used:
   Green 2 liters capacity Bottle-1
   Pink Water 1 liter capacity bottles-2
   Transparent 2 liters capacity Bottles-2

Fig. 1.1

c) Clean them with warm water and detergent. Warm water helps clean all the unwanted liquid stuck inside the bottle
d) Remove the plastic label sticking on the outside of the bottle.
e) Scrub off any adhesive left sticking to the bottle.

Fig 1.2

Fig 1.3

Fig 1.4

Step-2

Make a Mark

a) Identify where you want to cut a base for the Lamp shade Using a marker, draw a line around the green bottle, as shown in Fig. 8

Fig. 2.1 **Fig. 2.2**

b) With a knife, cut a slit on this line. The cutter should be sharp enough to smoothly cut through the bottle.

Fig. 2.3 **Fig. 2.4**

c) Using your scissors, pierce through the slit you made with the knife and start cutting off the top. (You could also just continue to cut with the knife but I like to use the scissors for safety.)

Fig. 2.5 Cutting With Scissors.

Fig. 2.6 Cutting with a cutter.

d) After cutting, separate out the top from the bottom.

Fig. 2.7 Seprated top of the bottle.

Fig. 2.8 Separated top of bottle.

Step-3

Making a stencil for the Flower Petal Shapes

a) Decide the shape and size of the flower petals.

b) Ideally a 4" × 4" square piece of plastic, cut out from a plastic bottle, is good enough to make a stencil for flower petal shape.
(Make sure you do not hurt your fingers while using the sharp blades!!)

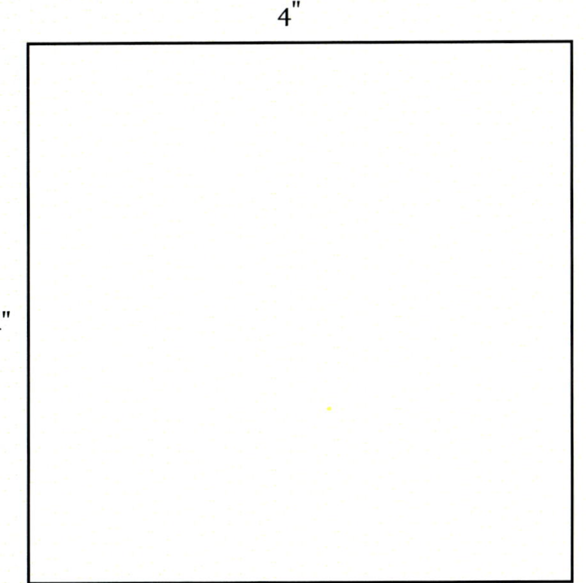

Fig-3.1

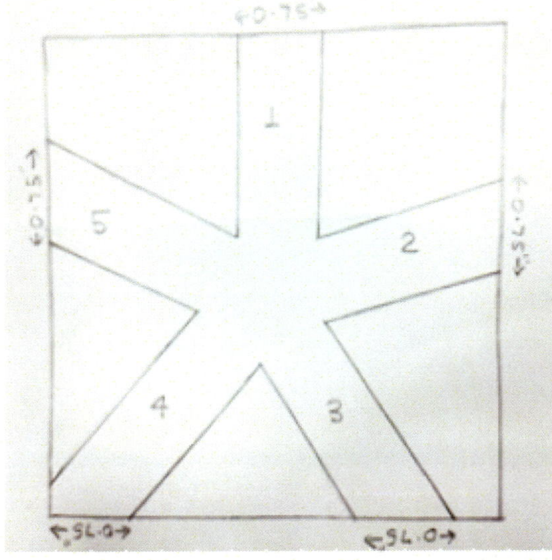

Fig-3.2

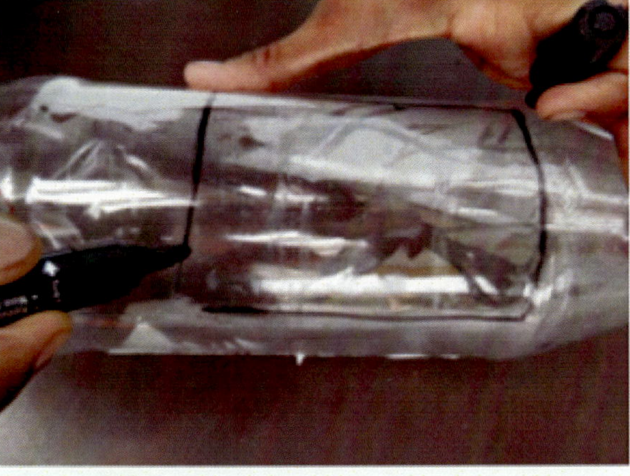

Fig. 3.3 Marking the piece to be cutfrom the transparent bottle.

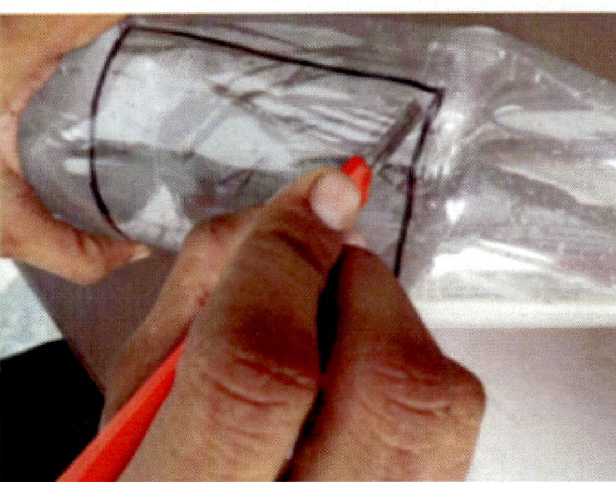

Fig. 3.4 Cutting the piece with a cutter.

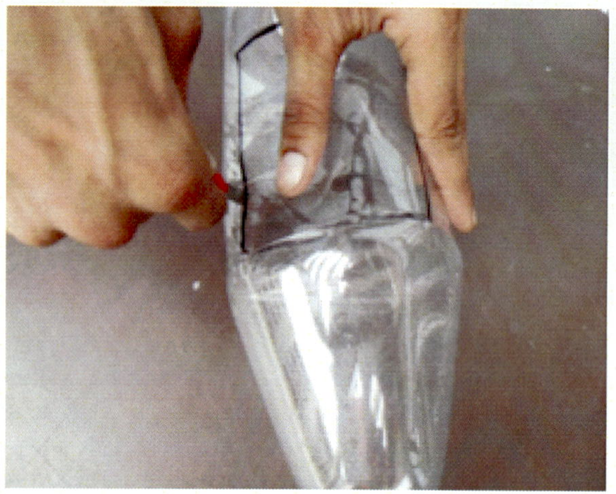

Fig. 3.5 Cutting out the Plastic piece

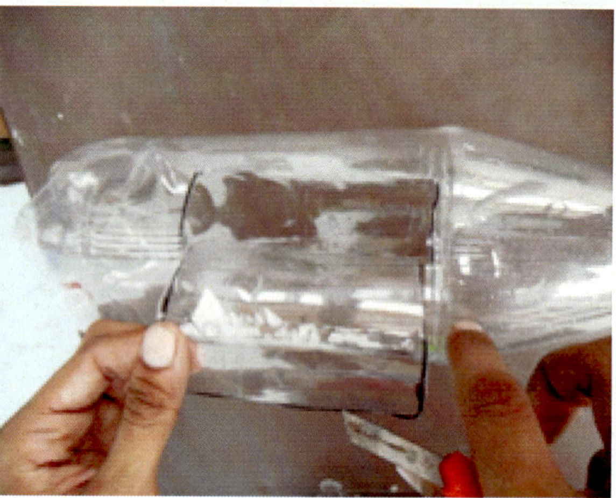

Fig. 3.6 Seperating the piece from the bottle

Fig. 3.7 Seperated piece for stencil for flower

Fig. 3.8

c) Draw the desired shape of the flower using a permanent marker on the 4"x4"piece of plastic.
d) Cut around this flower shape with the help of cutter/scissor.

Fig. 3.9

Fig. 3.10

Fig. 3.11

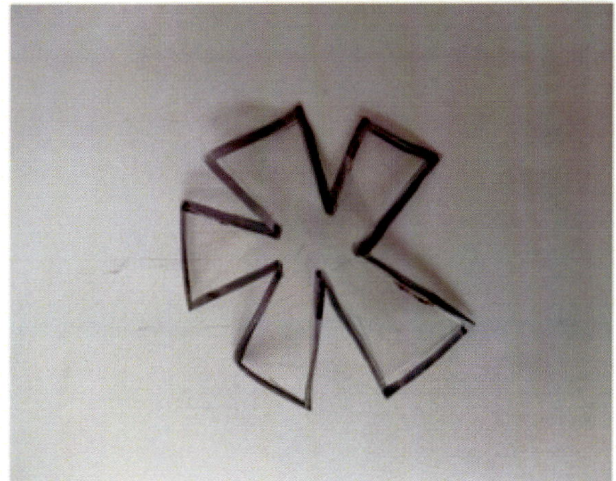

Fig. 3.12

Fig. 3.13 Final cut out and molded flower

Step-4

Mould It

a) Hold this basic shape of the flower over the candle flame.

b) Allow each petal to get heated up and softened.

Fig. 4.1

Fig. 4.2

c) With the help of tweezers mould each petal into a soft outward curvy petal.

Fig. 4.3

Fig. 4.4

Fig-4.5 Continue to shape the petals into the desirable shape using tweezers.

d) Make 25 flowers in this manner.

Fig. 4.6 Molded Flower-Final Shape.

Step-5
Filaments for the Flowers

a) Cut a 2"x 4" piece of plastic from the transparent bottle.

Fig. 5.1

b) Mark lines on this piece at a distance of 2 cm each
c) Cut out these narrow pieces of plastic.
d) Bring each of these strips close to a candle flame and as it gets heated up and softens, curl it up using a tweezer.

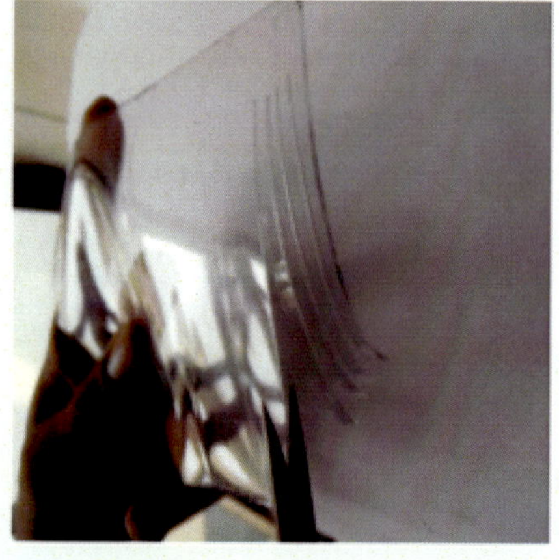

Fig. 5.2 Cutting the strips.

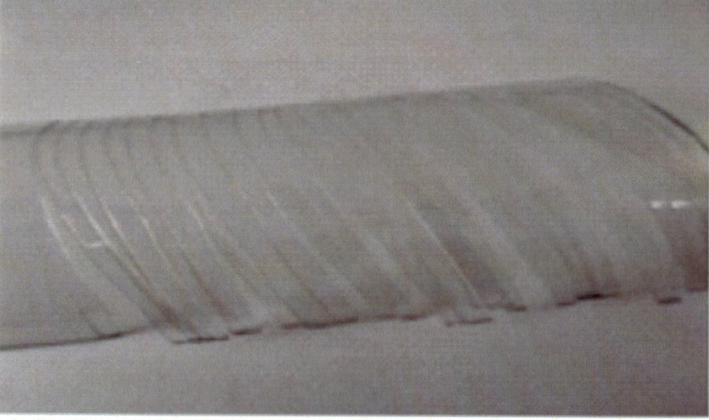

Fig. 5.3 Cut out plastic strips.

Fig. 5.4 Rotating the plastic filaments.

Fig. 5.5 Alining the filaments.

Fig. 5.6 Moulding the filament.

Fig. 5.7 Molded filaments.

Make one filament bunch for each flower. Hold them together in a bunch, heat them at top and base; Use a tweezer to do this.

Step-6

Stick these filaments to the center of the flower using fevibond stick.

Fig. 6.1

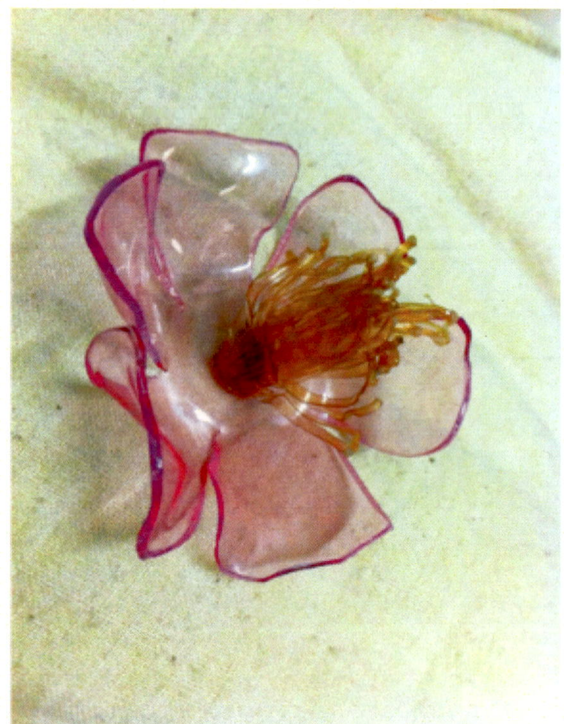

Fig. 6.2 Ready Flower

Step-7

a) Cut a strip of size 0.5 cms × 20 cms from the bottle to make a stem for the flower.

Fig. 7.1 Molded Stem

b) Bring this piece close to the candle flame and as it softens, mould it into a curvaceous shape as shown in the picture. (Make more stems and keep them for later use to make other products)

Step-8

Now we need to stick all the flowers on a green PET bottle with the help of Fevibond/ Fevikwick.

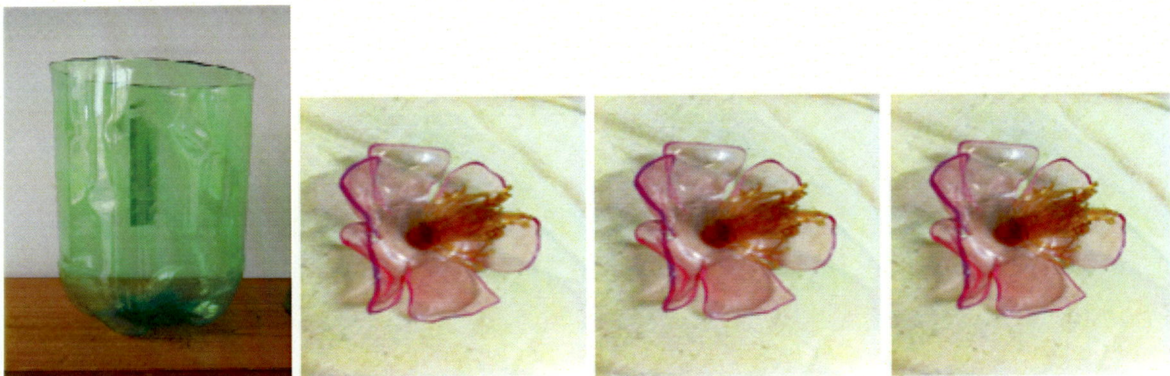

Fig. 8.1 **Fig. 8.2**

Fig. 8.3 Attaching flowers to the green PET bottle

Step-9

a) Take a holder with a base and fix it with a bulb.
b) Place the lamp shade, made above, on to this holder bulb with open side down.
c) Connect it to a socket and let it light up your favorite space.

Fig. 9.1 Placing the lamp on the holder

Fig. 9.2 The Personal Lamp is ready

The beautiful lampshade is ready to be used and gifted to loved ones or to people who value the future of Mother Earth.

Fig. 9.3 Lamp glowing in the Dark

Other Alternatives-Pretty looking stem of flowers–can be given to Delegates at a Conference The stem made in step 7 can be used to hold the beautiful flowers made in the project.